"

What Readers Say. . .

The approach taken by author, Fran Szarnicki, was one of the best that I have ever read. She was able to clarify in simple terms, the importance and relevance of numbers to your life and living every day. Prior to reading her work I really had no understanding of what numerology was." - Kay D.

"I have always wondered (about) the meaning of numbers in my life. My Dad was a numbers man and passed his fascination of what they mean to me. I am so glad that I purchased this book on Numerology by Fran Szarnicki as it has helped clarify so many questions about the effect that numbers have on a person's life. If you are thinking about purchasing a book to answer any questions about the effect of numbers you should definitely purchase this book." -CG

"I've always had an interest in numerology. The sequence of this book is easy to follow. And the charts help my memory." Mary B.

"This book is so interesting and was hard to put down! I really enjoyed her insights and personal stories to illustrate her points. Absolutely fascinating and very well written!" - Chris C.

(From Reviews on Amazon.com)

Numerology:

A Key to Understanding the Language of Numbers

by

Fran Szarnicki

Copyright

Cover/Interior Design: Paula F. Howard/A Howard Activity, LLC

Published by Aha!Press, an imprint of A Howard Activity, LLC

ISBN: 978-1-7362061-8-8

Printed and bound in the United States of America

Dedication

To my grandsons
Henry and William Rau

As two brilliant, philosophical, mathematical boys,
I believe you have a destiny to further mankind's
appreciation of Numerology.

Table of Contents

Prologue

I give thanks for the information that is in this book. I open myself to receive the knowledge that comes to me. -Francesca Szarnicki

Numerology sparked a passion in me that has endured since I began my study many years ago. When first introduced to the subject, I remember being consumed with fascination, and I devoted countless hours of study and reflection on the fundamental principles.

My love of this esoteric philosophy evolved into a life long quest, and the meaning of numbers became woven into the fabric of my life. The subject continues to delight, astonish, and even boggle my mind because it so easily conveys an understanding of people and their complex interactions with one another. I deeply appreciate how numerology is able to expand the awareness of the most serious and logical scholars. This knowledge provides a special key that can unlock the mind and move a person to a new level of comprehension and satisfaction. The change of perspective that it affords gives one a sense of freedom from the limited parameters that society routinely prescribes. Over the years, it has often caused me to exclaim, "Aha! Now I understand."

Being born on the ninth day of the month, I am naturally inclined to want to share the blessings of numerology with every person on the planet. (The idealism projected by my nine birthday endows me with this proclivity.) For years I have been experimenting with different formats in an attempt to relay this body of knowledge to others. It has been no small task to organize the vast amounts of data that I have collected so that people can easily understand the information and make practical use of it.

My first attempts at sharing my expertise began by offering my services to the continuing education departments at local colleges and universities. The numerology classes sparked interest in the community and before long a popular radio host invited me to be a guest on his show to talk about the subject. That evening program stimulated interest in a greater audience. Phone lines were always lit, and I soon became a "regular" feature on the show. The positive reception I received on the radio motivated me to look for more ways to address the public.

Next, I joined a speaker's bureau which provided the structure and connections I needed at the time. The agency created an outlet for me to speak to many people at meetings, conventions, and other gatherings. It became easy and natural for me to step up to the podium. I thoroughly enjoyed entertaining others in the process of teaching numerology. While I continued my speaking engagements, I found that although my audiences enjoyed learning about the subject, they were also quite busy managing their own careers. They wanted the personal information that numerology provided but preferred someone else to read their numbers for them.

After much thought into the matter, I decided to create a business combining my talent for entertaining with my numerology skills. I named this new enterprise Francesca after myself, and it grew to become a smashing success. I had found a recipe that perfectly satisfied my audience.

During personal consultations, it became commonplace to have clients ask, "How do you know these things about me?" I have had some people accuse me of calling the hospital where they were born to find their time of birth. They believed I was doing some kind of fabulous astrological research concerning their characters. Clients have asked me if I called their parents or spouses for information concerning certain events in their lives. I have been envied and congratulated for having special powers. Psychics have told me that my readings were the best they ever had. The truth is that a host of information concerning love, money, marriage, education, career and health about each and every person is written in their numerology.

While I have helped many individuals through my Francesca business, I cannot avoid the responsibility I feel to make this knowledge available to the general public. As I continue to be enriched by the study of numerology, I again feel motivated to share my unique insights with others. I finally decided a book would provide the best medium to reveal how a person can learn to read the language of numbers.

Although many beginner texts on this subject already exist, I realized that my understanding of the elementary building blocks in this field presents a slightly different point of view. This primer is the first in a series of books presenting my perspective on numerology. Each progressive lesson flows naturally from the preceding one making it very easy to learn. You will find that becoming number literate is no more difficult than learning to read the letters of the alphabet.

Everyone knows that very young students can and do develop reading skills. Based on our ease of reading the alphabet, I believe that even an elementary school student is capable of acquiring an understanding of the association of number symbols to specific meanings. By following the basic lessons presented in this book, virtually anyone can learn to read numbers.

My gift to you as I share my perspective of this ancient language of numbers is the freedom from the tedium of memorizing random facts about each number. I have presented the material in a way that allows you to absorb the knowledge with minimal effort. Soon you will begin reading numbers as if they were the letters of the alphabet. If you are already a numerologist, I hope that you will enjoy my viewpoint, and that my work will enhance your work.

Welcome to
Mrs. Szarnicki's Arithmetic Class

As we begin our esoteric arithmetic class, I want to point out that all of us are numerologists. You may not be consciously aware of it, but you constantly decode numbers. As we take a closer look, you will see that while we were learning to use simple arithmetic functions in grade school, we also inadvertently acquired the ability to read numbers.

This skill was quietly stowed away and carried with us into adulthood. The adult mind subtly treats the ciphers of the number set as the letters of another common alphabet. Whether we experience this knowledge consciously or unconsciously depends on our level of awareness.

For some people, the idea of unconsciously using the number set as an alphabetical system is a thought just too incredulous to be taken seriously. The fact that they never received formal instruction from their teachers prompts them to insist upon the impossibility of reading numbers. These people protest and proclaim that numbers were never meant to function as letters.

The argument goes something like this: Two dancers plus two dancers equals four dancers. Numbers are about quantities and they cannot tell us if the dancers would prefer a waltz or the tango. Numbers certainly don't indicate anything about the dancers' aptitude on the dance floor!

As we take a critical look at numbers and their qualitative meanings, I only ask that you keep an open mind. Perhaps one day you will be using numbers to decipher the potential ability of the dancers in the above equation.

To help you understand that numbers are *not* used strictly to indicate quantity, we will approach the subject with what we *do* know. Counting on the fingers is an easy introduction to the quantitative meanings of the numbers. The meaning of "how much?" or "how many?" can be visually displayed by holding up fingers on our hands. This is what the preschooler does when he embarks on his first lessons in math. We speculate that this is also how prehistoric man once communicated information concerning how many animals could be found roaming outside the cave. Humanity has been exposed to working with numbers in this fashion for eons. In recent times, we certainly have become tremendously more sophisticated in our ability to perform mathematical functions with numbers. Our group consciousness, however, continues to focus on the quantitative use of numbers just as we did in prehistoric times.

Most of us were not instructed to think outside the arithmetic box; therefore, we didn't explore the qualitative aspect of the number symbols while we were growing up. Our teachers never informed us of the flip side to the math coin because they probably weren't aware of its existence. Had they known, the door may have opened to some very enchanting information.

Early in my career I worked as a substitute teacher. One morning a local school district called me to handle some algebra classes. This particular call made me shudder. All I could envision was a bunch of rowdy teenagers, laughing, talking, throwing paper wads and generally ignoring anything I would possibly say. I reminded the secretary making the call that algebra was not my specialty, however, the district was desperate that morning and she had no one else on her roster to call. Before agreeing to the job, I asked if the administration would have any objections to me giving a lecture on the hidden meaning of numbers. They had no objection. In fact, they gave me free reign to talk about anything I wanted to, as long as that the classes were covered.

So I packed my numerology notes into my brief case and drove to the high school that morning prepared to address the algebra classes. As the late bell rang, I found myself standing in front of the black board facing twenty-six teenagers. This group had never met me before so I knew that they would study me for a minute or two before they started their party. I used that brief time to introduce myself as their special guest speaker and I immediately launched into my numerology lecture.

As I continued to talk, I could see, by the expressions on their faces, that I had captured their full attention. The students listened carefully and followed every word that I had to say. They were completely silent as I spoke for forty-five minutes! When the bell rang signaling the change of class, to my amazement, I heard students objecting, "No! No! This is not fair. We are not leaving until you tell us more." One boy refused to get out of his seat until he was overrun by the next class filing through the door.

In my algebra classes that day, I shared the hidden truth that numbers are codes of a secret and ancient language. Numbers are not just symbols to indicate amounts. They were not invented solely to do mathematical functions. Numbers form an ancient coded language that beckons to be read. This language is a physical part of all of us. We each have one brain with two hemispheres. We have five physical senses. We have two hands with five fingers on each hand. For eons every civilization on this planet has used a two hand counting system. The language of numbers you see is not a foreign language. It lives as a component of our minds and our bodies. Because it is an integral part of our being, it can help us to understand our existence and unfold our spiritual purpose. Numbers make sense of everything.

The language of numbers is something that you are already using in your daily life. You constantly make interpretations and judgments, based on the codes that each and every number contains. Everyone does this both consciously and unconsciously. This language is our steering mechanism as we travel along our life paths. To demonstrate your excellent number reading ability, we will compare two story formats that are exactly the same; however, we will plug in different numbers into each format.

FORMAT A

My name is Wanda. I live in an apartment at 1000 Fifth Avenue in My City, USA. I am the proud 1st owner of a hot little car. I have a hard time keeping the speed under 80 mph. This morning I left the gym early because my weight was finally under 125lbs. I went to my class and found out that I got a 98 on my last accounting test. It was a relief to know that my QPA will not drop below 3.86 this term. My goal is to be able to make $250,000 per year when I graduate.

FORMAT B

My name is Wanda. I live in an apartment at 16½ 4th Avenue in My City, USA. I am the proud 3rd owner of a hot little car. I have a hard time keeping the speed under 25 mph. This morning I left the gym early because my weight was finally under 200lbs. I went to my class and found out that I got a 76 on my last accounting test. It was a relief to know that my QPA will not drop below 2.0 this term. My goal is to be able to make $25,000 per year when I graduate.

COMMENTS ON THE STORY FORMATS

- *I find it a little disturbing that Wanda #2 lives at an address that is only 2 digits. Most addresses usually have 3 numbers. The ½ is unsettling. I don't know if I would want to visit Wanda #2 at her apartment.*

- *I bet that Wanda #1 has a really swank apartment on Fifth Ave. It seems that the Fifth Avenue of every city is the place to be. Fifth is usually where you can find the vibrancy of the city. I bet Wanda#1 has a hot body to go with her hot car.*

- *Wanda #2 should have stayed at the gym longer and studied harder. Because Wanda #2 has not set very high goals, I am guessing that she may have a low self-esteem issue.*

- *Wanda #1 is a smart cookie.*

As you read through the story formats, were you making judgements? Did you conjure up images of the two Wandas? Were you reading numbers? As you encountered the figures showing the weight of the two women, those amounts may have stimulated a picture to form in your imagination. The street numbers don't indicate amounts, but they also paint a picture. Most peoples' perceptions change when we say Fifth Avenue, Fourth Avenue, or Second Avenue. Why is there a tendency to think of Fifth Avenue as a busy, lively and happening kind of street? How does this commonality in our thinking pervade reality?

In the case of a street name, the mood is evoked by nothing other than the actual number assigned to it. Perhaps, merely the presence of the number Five creates an electric atmosphere along Fifth Avenue. Is there a kind of mysterious mood that emerges when we name a street Two? If I ask a taxi driver to take me to Second Avenue, does that driver have to be afraid? Is Second Avenue traditionally a place where shady dealings take place? What goes on behind the closed doors along Second Avenue?

There is a popular cliche that states: *the third time is the charm.* We smile and tell ourselves that we will finally be successful (or lucky) because we have repeated our attempts three times. Why is this so?

Retailers know that customers tend to buy when a price ends in a nine (ie: 29, 39, 49, etc.). Logically we know that .99 cents is only one penny away from one dollar but it has been statistically verified that we are more likely to make a purchase when we see .99. Could it be that there is something about seeing the nine that causes us to part with our money? Gestation took nine months and then we parted from our mother's womb. Could nine be a force that causes departure?

I ask you to think about these things with an open mind. Don't be a square. A square has four sides, so I'm really requesting you not to think like a number Four. I want you to think outside the box. Have you ever known someone who was a square? Squares (Fours) do everything according to the rules. They are very repetitive and traditional in thoughts and actions. Although I appreciate tradition for the stabilizing influence it provides, I need you to leave that security now, and prepare to explore the language of numbers.

Using the Key

In the following pages we will examine each of the numbers, plus zero. The numbers One through Nine are presented from two distinct vantage points. Each explanation of the numerical concepts contains two sections entitled **THE PHILOSOPHY OF** (the number) and **THE HUMANITY OF** (the number).

The philosophy section takes you on an investigation of the general meaning of each number. As you absorb the significance of these meanings, you will find yourself becoming familiar with the realities constructed by these numbers. Distinct moods, feelings and vibrations emanate from each numeral. Some of the descriptions are more abstract than others because certain numbers have a natural tendency to be more ethereal. The philosophy sections can be quite intense; therefore, it was my intention to keep these segments brief.

When reading through the philosophies, you will find that each account is contingent upon the previous descriptive reality. As I introduce each new philosophy, I begin by reviewing the properties of the preceding numbers. By weaving the concepts together in this way, you are compelled to re-examine and remember the meaning of each number. All of the philosophy sections have a variety of quick reference guides at the end. These easy to understand capsules include illustrations, descriptive adjectives, job placement suggestions, first person comments, biblical references and other miscellaneous associations.

ILLUSTRATIONS graphically presents the symbolic meaning of each number. **DESCRIPTIVE ADJECTIVES** give one word summations that capture the essence of a particular number. Similarly, the **JOB PLACEMENT** category offers careers that portray the characteristics of a particular number. The category entitled **CONTEMPLATE** gives you the opportunity to reflect on a passage from the bible where there are often direct references to the number.

In the category entitled **(1,2,3,4,5,6,7,8,9) SPEAKS** we have numbers expressing their values directly to us in their own words. You will find that throughout my writing I frequently refer to the numbers as entities. In any given passage, a number can be a he, she or it. I use these pronouns interchangeably to explain numerology concepts. Hopefully, this referencing will further enhance your ability to relate to numbers as living forces. I want you to understand that each number has a unique personality and tenders a special relationship with us. The final capsule offered at the end of the philosophy sections entitled A **GLANCE AT (1,2,3,4,5,6,7,8,9)** is just that—a brief scan that offers a variety of descriptive associations with a particular number. Do not be overly analytical about the categories and the models presented here. For example, Leo is offered as a zodiac sign to represent number One. This does not mean that other zodiac signs cannot contain One energy. When you skim the column (Helios, Leo, Masculine, Conscious, Mental, Red, Fire, Aggressive) this capsule should give you an instant flavor of the number One. That is the purpose of the **A GLANCE AT** category.

A section entitled **THE HUMANITY OF** follows each number review. These segments differ from the Philosophies as they offer real life examples of people who display notable characteristics of given numbers. These case studies illustrate and reinforce the information presented in the philosophy section. The descriptions allow you to see how the philosophy becomes animated and plays a genuine role in peoples' lives.

In a few of the reviews, the influence of the numbers is derived primarily from the sum of the letters/numbers in the name. Many of the examples in the Humanities section, however, reference several different influences found in the complete numerology chart. To protect their privacy, I have changed the names of the people in my anecdotes.

Zero

DIVINITY

Zero, the cosmic egg, represents the potential of divine energy, and it represents nothing at all. If we are told that there are Zero cookies in the cookie jar, we know that there are none to be found because Zero means none. It indicates a state of emptiness. As we contemplate the emptiness of zero, let's direct our attention to the void of space. For ages, classroom textbooks have referred to space as being vacuous. As scientists continue to explore this notion, they are realizing that space is not empty at all, but filled with energy. Without the stored invisible energy in space, nothing in the visible world would exist. Scientists are finding that the nothing of space is really something, and that something (nothing) is the basis of all that is. This dynamic paradox is recorded with the zero.

It's quite amazing to consider that the reality of every action occuring on the physical plane derives energy from the stillness of the universe. In reviewing the details of the lives of highly accomplished people in diverse cultures, we find records that lead us to believe that many of those people purposely charged themselves with stillness. Many of the great masters of the arts and sciences incorporated a meditation practice into their daily lives to facilitate their success. Thomas Edison, for example, did much of his deep thinking and retrieved many of his most powerful creative thoughts while sitting in the darkness holding a handful of marbles. He created this practice to allow himself to go into an altered state. The purpose of holding the marbles was to keep his mind minimally aware and conscious. A part of him had to stay present so he would not drop them. Maintaining this bit of awareness was sufficient to bring his deepest inventive thinking back to the surface.

Whether or not we are aware of our need, we all crave the experience of transcendental consciousness. Because it takes only a few fleeting seconds in the transcendent to receive the special nourishment this state provides, the process of absorption usually continues unnoticed. We soak up the bare minimum as we transition between the waking and sleeping states of consciousness. When we fall asleep and as we awaken, we pass through a pivot point where we are neither asleep nor awake. After having a clear and conscious episode of awareness in the transcendental state, many people seek to repeat this intriguing experience. Practicing meditation often proves to be a reliable approach to creating the proper atmosphere so more of these events can surface.

Think about the transcendent in mathematical terms. If plus one equals being awake, then minus one would equal being asleep. If you are neither asleep, nor awake, then where are you? You would be nowhere, neither here nor there. You can't put your finger on nowhere because nowhere is everywhere. Zero represents this place called nowhere, and remember, nowhere is everywhere.

There is no time in nowhere, therefore zero also represents timelessness. Timelessness is the place where we shift the mind beyond our current reality. This state is commonly referred to as the transcendent. Timelessness creates the ***I AM*** experience, or the direct experience of God. When we infuse timelessness into our individual beings, we are filling ourselves with the universal life force.

We know that Zero indicates power because it stands for the energy called God. If we equate God with love, then we can conclude that Zero embodies God's love. What a blessing it is to have nothing. In having nothing, you have the world. In having the world you have it all. We begin with zero, with nothing, because nothing provides the basis for everything. How lucky we feel when presented with the opportunity to have a fresh start. God's love and forgiveness allows us to erase the slate and prepare for a new beginning.

Zero is not a number, but it has the power to shift the energy of any number. Zero can turn a 1 into a 10 or a 100. The introduction of the Zero greatly changes the meaning of the number One. There is quite a range of difference in receiving a one dollar bill, a ten dollar bill, or a one hundred dollar bill as a gift. Each added Zero shifts the energy and meaning of the number, making it increasingly more valuable. (Remember, nothing is everything.) Unbounded potential exists in the little cosmic egg called Zero.

The Philosophy of One

UNITY

When Zero, represented by the circle, the symbol of God, opens, unbounded transcendental energy makes itself known. As the ***I AM THAT I AM*** imprints itself in a single conscious mind, it instantaneously creates a presence throughout this third dimensional reality. I refer to this miraculous occurrence as *"the hatching of the cosmic egg."* As a result of this event, the number One comes into being.

The number One is a unified force just like that of the God energy, and it contains more potential than could possibly be used in a single lifetime. As the masculine presence of God, it emits the brilliant and self-reliant energy of a first born son. As the figurative son of the father, the One exudes a profound confidence that originates in the dominant position it holds. The fearless One inherently knows that it was born to lead and views challenges to its command unkindly. One gives orders with authority and expects greatness from others. Its keen self-awareness keeps the One attuned to an unlimited creative power source--itself. One exemplifies an extraordinary person, and those who follow the One are likely to model their behavior after this unique individual.

It is the nature of the One to be assertive, and its eagerness to take off running constantly shows. One has a strong pioneering spirit and comfortably challenges uncharted territory every chance it gets. It is quite an amazing spectacle to witness the creative flow of the number One energy unfolding. An overabundance of confidence enables it to sail effortlessly through its undertakings the very first time. One can be busy hatching newly conceived plans, and at the same time work to convince onlookers that it is highly experienced at the many tasks that lie ahead.

The unity principle does not permit the One energy to be diluted; therefore, the intensity of One's energy on the creative stage is always quite compact. A central part of fulfilling the role of being number One involves planning an entrance and designing methods to maintain its inventions. Although it investigates with the ease and innocence of a child, it knows that developing an infrastructure is equally important. After conquering new territory, One quickly refocuses attention to constructing a path for others to follow. As soon as it completes this essential task and the venture can stand on its own, the accomplished One quickly readies itself to tackle new challenges. It gladly leaves the day to day routine maintenance chores to others while it moves on to explore and conquer in a new field of opportunity.

The investigative spirit of One naturally anticipates finding more of its own inherent strength and perfection reflected in its world. This symbol of unity holds high expectations because it embodies the principle of excellence. Stubbornly refusing to settle for anything less than the best, One often becomes annoyed when encountering imperfection.

When One makes a purchase, it expects the merchandise to be flawless, and refuses to consider inferior goods. Whether out on a leisurely shopping excursion or solving the world's problems, One has difficulty understanding how anyone could possibly compromise his or her standards. The supreme power One commands is a direct result of its unified thinking.

MOTTO: NEVER COMPROMISE

ILLUSTRATIONS

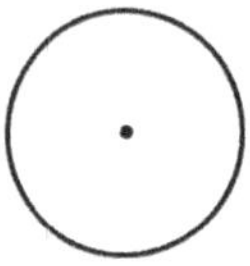

1.

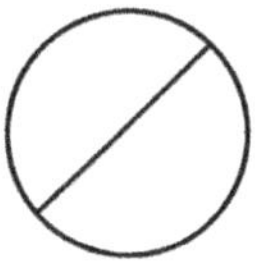

2.

1

3.

1. The dot in the center of the circle indicates life emerging from the cosmic egg.
2. The slash breaks open the cosmic egg.
3. Modern day symbol for one

DESCRIPTIVE ADJECTIVES

Independent, Courageous, Assertive, Creative, Bold, Dynamic, Spirited, Achiever, Adventurous, Energetic, Inventive, Confident, Ambitious, Determined, Commanding, Influential, Forceful, Enterprising, Original, Authoritative

ONE SPEAKS

- There isn't anything that I cannot do.
- I expect excellence, and excuses do not sit well with me.
- I'm number one, and don't you forget it!

JOB PLACEMENT

Executive, Speaker, Inventor, Designer, Engineer, Business Owner, Department Chair, Manager, Administrator, President, Artist, Sales, Model, Talk Show Host, Guide, Writer, Principal, Composer, Architect, Team Captain, Director of Quality Control

CONTEMPLATE

The Story of Creation / Day 1

God said "Let there be light," and there was light.

Note: Every living cell contains light. Cells generate their energy and chemistry from inherent light frequencies. In essence, life as we know it would not exist without this initiation of light.

A GLANCE AT ONE

Mythology Figure:	Helios or Sol, god of the sun
Zodiac Sign	Leo
Gender:	Masculine
Mind:	Conscious
Temperament:	Mental
Color:	Red
Element:	Fire
Energy:	Aggressive

The Humanity of One

A Unique Individual

My daughter, Lynne, has an unprecedented amount of One energy flowing through her being. She began to assert her independent and adventurous spirit at a very young age. This fearless baby was very clear about what she liked and disliked. She rarely asked for assistance when she went exploring. Determination empowered her to succeed in her endeavors. I remember the times when she was still a preschooler and we would go out shopping together. I would follow behind her whenever possible to allow her abundant One energy the space and freedom to lead the way. I'm sure it appeared odd to people to see such a young child pushing a cart and shopping on her own and I received many glaring looks of disapproval as I trailed behind. Knowing my young daughter had a great amount of One energy to express made me extra sensitive to her needs.

Lynne has always had the One competitive spirit. She often makes a game of turning situations into a strategic operation with the goal being "First Place". For many years she did battle on the golf course. During college, she won the individual division title for her school four years in a row and was named the MVP in women's golf each year. She became the player to beat and that tag only gave her more fuel to prove herself. Lynne describes her competitive spirit in this way:

It is a driving force that won't let go of me. I recently went bowling with a group of friends. I'm not a big fan of bowling; however, we were just out for a night of entertainment. On this particular evening, I could feel the aggression building. I have a need to win that often wells up in me when a challenge is presented. I start feeling the grrr in my jaw, and the next thing I know, I am saying things to cause my friends to make them doubt their ability at this silly game of bowling. I go into a mode that drives me to outwit the opponent at all costs. I am learning to control this crazy competitive feature of mine. I do restrain it at times, but it is always there.

Academically, Lynne pushed herself to achieve and graduated with honors. She explained to me that because our last name begins with an SZ she knew that she would be doomed to be one of the last in line to receive her diploma. That, of course, was not acceptable, and the only way she could find to move herself to the front of the line was to study hard and graduate with honors.

Her need to be outstanding extends itself into every choice she makes. When we are out shopping, she is careful with her selection. Every item must be scanned for imperfections before making a purchase. The fit and color of all clothing must accurately flatter her image. As a result, everything that hangs in her wardrobe is perfect and absolutely suited for her. She never participated in swapping outfits for special occasions. These days, she and I wear the same size, but I know better than to ask to borrow anything. Her clothes are stamped with her personal cosmic identity. If I were to wear something that belonged to Lynne, she would feel that it was no longer hers exclusively, and she would never feel comfortable wearing that piece of clothing again.

Today, Lynne and I have a very close relationship. We understand each other very well not only because of the experiences we have shared, but because we are both number literate. Studying numerology has helped us to support one other in ways that we might overlook without this means. For example, when we are traveling together, I automatically expect her to take the lead or drive the car. She is always the person in charge unless she delegates that responsibility to me. I sometimes jokingly say to her that *I would never knowingly walk in front of a number One person*, and she will usually comment back to me that it is not really necessary to go to such extremes. (Lynne does have an occasional weak moment of wanting to be polite and reasonable.) However, she knows deep inside that if I were to command the lead, her spirit would be incensed. Raising a number One child has taught me to respect and accept the gift all trail blazers give to the world. They were born to initiate action and take charge.

There are many people in the world who need someone to guide them, and Lynne attracts those people because she has an insatiable urge to take the lead. In light of this energetic configuration, she functions well as "the boss." Lynne was 22 years old when the directive One energy dared her to go solo with a business venture of her own. Today, she owns a food truck and supplies grocery stores with her products. Being a business owner allows her the freedom to be as creative and bossy as she pleases. She makes it clear to others that as *the woman in charge,* she will not be challenged. Plans cannot be forced upon her. Her headstrong nature becomes wary if she suspects someone of trying to manipulate her dreams. Although her inventive nature supplies her with a well of unique solutions to everyday problems, Lynne does accept helpful suggestions from her employees. She periodically sorts through their input; however, the decision to implement a select idea is reserved for her alone. Because she is secure and comfortable with her position, she is able to incorporate the best suggestions from a variety of sources. In this way, her vibrant and unique food business continues to grow.

I have often warned my daughter and others who carry the strong One energy that they are bound to attract weak people. Because they are experts at barking out orders, Ones need to have people around them who are willing to listen and accept directives. On occasion, Ones will attract like energies, and they enjoy being with other creative powerhouses. In these situations, however, it is important that the Ones honor each others boundaries. It only makes sense that friends who are both number One's not attempt to command the same territory.

Amy, Alice, & Andy

Amy, Alice, and Andy are siblings who all display strong number One characteristics. Amy wields the most power in the group. To a numerologist, it's obvious that the A, the first letter of the alphabet leads the way for these individuals. Amy's A has a hard pronunciation as opposed to the softer sound in Alice, and Andy. This strong intonation of the first letter of the alphabet firmly connects her with the power of One. When deciding on a college, Amy chose a small academic institution and claimed English for her major. This decision was influenced by the fact that if she had chosen a large university, she would have risked becoming lost in the crowd. Since she embodies strong One energy, it was important to her to stand out among her peers. Amy was very attentive to polishing her talents so she could become the big fish in the small pond. She achieved academic excellence by honing her writing skills. After graduation, her need to lead was fulfilled when she took a position as an administrator of a cyber school.

When it was Alice's turn to go to college, she felt that she had to establish herself in an environment uniquely different from that which her sister had chosen. As a result, she opted for a university with a large urban campus. Alice has superb writing skills, and her sister, Amy, felt that she would have excelled by making English or Journalism her major. Alice's number One style of thinking, however, urged her to explore something different. Unwilling to share territory Amy had already conquered, Alice became an art history major. In her mind, taking a unique path would allow her to bypass potential problems. The last thing Alice wanted was to compete with her sister. Currently, Alice proves herself to be a unique and superior number One through her outstanding artistry. Although she seems to be appreciating her own writing skills more, she still refuses to take full advantage of her natural journalistic talent. Because of our interviews for this book, the sisters have been open to discussing the workings of the number One in their lives. Although they are now consciously aware of how the number One has influenced their decision making, the plans that each of them has set into motion continue to unfold accordingly. It will be interesting to see if the ruling power of the One is tempered over time with this new found awareness.

Andy, the baby of this family trio, is not shy about expressing his unique and compelling individuality. This number One boy has two powerful older sisters to outdo before he takes on the rest of the world. His tactic, so far, has been to overachieve in everything he does. He currently maintains straight A's in the arts and humanities as well as in math and science. Not only does he have better grades than his sisters did when they were in school, he also retains the status of being at the top of his class. His perceived need to fight for position has brought out the absolute best in this young man.

Andy's obsession with being "number one" is infused into every facet of his being. His need to stand out in the crowd becomes obvious when you see his attire. Andy has always attracted a lot of attention by wearing wild colored clothing. This fashionable, good looking, and popular fellow celebrates his unique style every spring when he struts out on the runway for annual fashion shows.

His personal favorite extracurricular activity is his involvement with music. Andy shines as the lead guitarist for the band he started as a young teen. He has written many of the songs the group plays, and he consistently works to develop more material. His primary goal for the band is to exclusively play his original work. With all of his tremendous accomplishments, Andy also manages to hold down an after-school job that carries a lot of responsibility. He works at a small business where he does the books after closing the cash drawer every evening. It seems the owner of the business has become dependent on this dynamic young man and when Andy does move on he will be sorely missed. The usual next step for a high school student with Andy's excellent academic credentials would be to attend college. In his case, because independence dominates Andy's personality, he may find success without the stamp of approval of a particular college.

Although it's common for competition to arise between young siblings, this issue is particularly intense between Amy, Alice, and Andy. The drive to out do one another clearly goes beyond a need for parental approval. Numerical coding in their names compels each one of these characters to **BE NUMBER ONE**.

The Philosophy of Two

RELATIONSHIP

While Number One initiates a process of discovery of the world in which it finds itself, in due time, however, the urge to explore the environment will come to an end. When One finally reaches a peak of satisfaction, the masculine force turns its instincts inward and begins to conduct a thorough investigation of itself. In order to become a true witness of the self, the monad must separate from itself, and this decisive action causes the number Two to come into being.

As a result of the trauma suffered during the division of self, the contemplative Two carries the sorrow of separation within its spirit. Uncomfortable with this parting, Two longs to return to the strength and unity of the One. The Two often feels great loneliness on its journey to know and understand the One. Nevertheless, it accepts the sacrifice of separation in order to further the cause of evolution.

Philosophers of past ages have symbolically used the element of water to describe how a man arrives at a point of self-realization. The sages relate the image of a man looking into a pool and seeing his own reflection for the first time. In the course of his self discovery, perhaps early man actually did identify with the image in the water. It is possible that this event triggered the initial steps in his self discovery.

We know, however, that the complex achievement of self-realization does not develop from gazing into a mirror. The quest for greater conscious awareness is a deliberate process that can be all consuming. A person usually moves through a full agenda of experiences, both internal and external, to fulfill this courageous mission. Experiences need to be processed, and it takes time to integrate the new expansion of awareness into one's being. Taking a careful inventory of the soul is an important part of getting to know oneself, but it is only the beginning of the journey.

The pilgrimage of self-discovery requires clear intention and commitment to the spiritual path. The number One takes on this challenge by dividing its spirit in order to objectively perceive itself. The One says *I am.* Two, the reflective energy, confirms, *Yes, you are.* This primary event is essentially how the One meets his *wo*man. As you will learn, the "w" is equivalent to Five (the number of man) and "o" equals Six (the number of the heart and the feminine aspect of man). Together the masculine and feminine vibrations become balanced and have the precious opportunity to realize wholeness. **The (*wo*) man is the balanced man.** The male can finally see himself through the eyes of his female self. It is the heart, or the (*wo*) man, or the number T(*wo*) that delivers the self-validation man sought when he divided.

It is apparent that the numbers One and Two are actually the male and female aspects of God. They represent two halves of one being and were never meant to function detached from one another. Deep sadness overcomes the heart when One and Two fall into discord. If these aspects become separated, they automatically begin searching for one another. Together this pair creates the experience of wholeness.

The Two is recognized as the female presence of God. In its highest spiritual ideal, the Two manifests as the principle of equilibrium. (The number two cannot be divided into two unequal parts.) The Two supports and balances the action of One. The human body exhibits multiple configurations of this principle. The body manifests unity with one brain, but the brain has two hemispheres with which to operate. We have one heart but it functions with left and right ventricles. We take one breath through two nostrils. We were designed to walk with two legs, rather than hop on one foot. True creativity requires the use of conscious and subconscious thinking. Both modalities must work together for optimal results in facilitating the ability to create knowledge. The balance of male and female energy in our society is absolutely critical and we become dysfunctional when it is disrupted.

We can draw a comparison of this principle of male/female equilibrium to the balance of power that we find in nations around the world. We recognize that in those countries where men honor and respect women as their equals, culture flourishes. In countries where men denigrate women, the system of government tends to be corrupt. This disparaging behavior eventually leads to instability and upheavals that in turn erode the culture.

A gentle and patient force, Two represents our inner

nature. The Two however, is not ego oriented. Like a chameleon, the Two adapts to the signals it receives from external influences as well as its higher self. Because it naturally seeks to cooperate with these powerful influences, Two often finds its place as a partner or an assistant. It always seeks compatibility within its environment and works hard to smooth out difficulties and cultivate peace. A strong inherent inclination to create harmony will often manifest musical talent in the Two giving it a desire to sing or play an instrument.

Finding satisfaction working behind the scenes, Two makes the ideal executive assistant. Two thinks deeply, then offers its best ideas to the person in charge hoping those concepts will be implemented effectively. Generous and kind, it continuously works to inspire leadership. It values the appreciation and acceptance received through partnership and does not seek to usurp power. Mutual trust between the Two and executive power can be maintained indefinitely as long as the boss responds with a positive attitude and some acknowledgement. The Two asks for nothing more than to receive the coveted approval and respect that it deserves.

ILLUSTRATIONS

1. . 2. 3.

1. The Roman numeral for Two shows equality and balance. The two side by side vertical bars emphasize reflection.
2. The half moon hovering yet attached to the earth (horizontal line).
3. The modern Two

DESCRIPTIVE ADJECTIVES

Agreeable, Modest, Friendly, Supportive, Courteous, Genteel, Emotional, Cautious, Considerate, Diplomatic, Hospitable, Receptive, Understanding, Patient, Kind, Harmonious, Sensitive, Sympathetic, Gracious, Cordial, Tactful, Humble

TWO SPEAKS

- Please, be my guest.
- We should find out what the dress code is for the party this weekend, so we can plan something appropriate to wear.
- Whatever you decide will be fine with me.

JOB PLACEMENT

Bank Teller, Arbitrator, Secretary, Homemaker, Librarian, Diplomat, Coach, Administrative Assistant, Auditor, Musician, Editor, Counselor, Psychic Medium, Seamstress, Illustrator, Psychologist, Dancer, Custodian, Systems Analyst, Aid

CONTEMPLATE

The Story of Creation / Day 2

God said, "Let there be a dome in the middle of the waters, to separate one body of water from the other." God made the dome, and separated the water above the dome from the water below it. God called the dome "the sky".

Note: There is only one body of water, but God separates it into two specific arenas. This passage calls attention to the idea that number Two functions separately, yet remains a part of the unified whole. There exists only a single body of humanity divided into male and female aspects.

A GLANCE AT TWO

Mythology Figure:	Luna, goddess of the moon
Zodiac:	Cancer
Tarot:	the high priestess
Gender:	feminine
Mind:	subconscious
Temperament:	emotional
Color:	blue
Element:	water
Energy:	passive

The Humanity of Two

The Diplomat

I found it interesting that when I began scanning my mind for examples of Two's feminine energy, almost all of the cases I could recall were men! My husband Jim, our son Stephen, and a good friend named Matthew are all *gentlemen* who truly live up to the title. All three of these men are peacemakers. I recently listened to Matt's mother as she reminisced about her son's boyhood. She said:

When Matthew was a child he always showed concern for other children in his school. He would check on those who were being bullied on the playground and made sure they were not harmed. He was a big kid so the troublemakers were not likely to challenge him. However, there did come a day when an incident flared. One of the agitators pushed Matt and challenged him to a fight. At the time, Matt didn't respond to this challenge. He walked away to think about the matter, as there were a number of options for him to consider.

The incident perplexed Matt so he waited until he got home to discuss what do about it with his mother. Since it is his nature to search for diplomatic ways to resolve issues, getting into a brawl was not his solution of choice. Matt needed to get an objective, adult opinion before he made up his mind on how he would handle the problem. After a conference with his Mom, Matt decided that the next time the aggressor challenged him with a shove, he would shove back. In due time, the bully was back to repeat his confrontation with a body slam into Matt. Matt reciprocated as planned. When the aggressor felt Matt's strength, he backed down and was never heard from again.

I am very familiar with the process the Two goes through before taking action. As natural diplomats, my husband and our son attempt to weigh all facets of a situation before making a decision. My guys are never quick to anger, and they usually retreat to the "strategy room" before getting involved in any aggressive conduct. If and when they decide to fight back, Jim and Stephen will assert their energies through the proper and civil channels. Questioning authority is something they both strive to avoid. Both men would rather pay an undeserved penalty than argue with the accounting department about an error on a statement. When either one recognizes that such a mistake has been made, he does whatever he can to avoid getting directly involved in the dispute. Jim and Stephen usually refer such oversights to someone else who is willing to solve the problem for them (oftentimes that is me).

Matt, Jim, and Stephen detest discordant situations and overwhelmingly refuse to be a part of that energy. Their natural inclination is to work to encourage harmony. One way that they have of promoting and integrating agreeable vibrations is through music. Jim is not one to call attention to himself, but he seems to loose his shyness when he has an opportunity to step on stage to sing. Some years ago, the church choir invited him to sing solo on Sundays during the communion service. Jim certainly knew he had the ability to perform long before they recognized his talent, but the introvert Two energy prevented him from promoting himself. Two never broadcasts talent but patiently waits to be recognized. When the choir extended the invitation, he accepted, and with his wonderful voice he infused peace into everyone who attended the mass. When we happen to meet old friends from the church, Jim still receives accolades for his solo performances.

It always amazed me how easily and naturally our son Stephen took to music. As a child, he seemed to automatically know how to read musical notation. Rhythmic talent manifested in his tap-dancing as well as playing the tuba in the high school marching band. He always participated in his school's annual musical productions and was never shy about taking on major rolls with lots of dialogue, singing, and dancing. Involvement in the musicals gave him an outlet for his natural talent and brought him tremendous satisfaction. Stephen's experiences nurtured a love and appreciation of music that will always remain with him.

Matthew is the one fellow in this group who is pursuing music professionally. He currently writes music and plays the guitar and piano. Matt sells his original material to screenwriters in Los Angeles and has recently produced his own CD. He recalls first seeing musical notation in his dreams when he was in high school. Although he went off to college with the intention of studying architecture, that fantasy faded as his natural inclination to become a musician grew stronger. The music of his dreams kept presenting itself until he could no longer ignore it, and he had to write it out for publication.

I know all three of these gentlemen to be sensitive and diplomatic in their approach to others. I recognize each one of them as a strong individual, yet not one of them is ever pushy when it comes to promoting himself. The attitude seems to be *I am willing to serve in a leadership capacity only if I am first invited to take up the task.* Because Jim, Stephen, and Matt all find rejection to be unbearable, they avoid putting themselves into situations where they might encounter it. They frequently display a passive attitude concerning positions in which their talent, skills, and training can be effectively utilized. In my opinion, their protective passive attitudes have too often prevented them from receiving the praise and recognition they deserve. I advise anyone influenced by the presence of strong Two energy to get an agent (or a partner) to promote them. People who carry an abundant amount of Two energy are not meant to tackle life's challenges alone. They function best when they are working in tandem with another person.

A Female Representation of the Two Vibration

I offer one more dramatic example of another very specific way the Two energy can effect a person. This feminine energy has aligned itself strongly with the physical body of a subject named Dianne. The direct effect of this placement manifests in her very attractive, petite, and delicate frame. Physically, her body demands constant attention, care, and protection. She exercises diligently but never appears robust. Even in her younger years, she was never been able to overextend herself by staying up late and partying with the crowd.

Extremely impressionable, Dianne seems overly sensitive to odors, and heavy perfumes can instantly make her sick. Her body can be quite reactive to different foods as well as drinking water to which she is not accustomed. It is the nature of the Two to be sensitive and receptive; therefore it logically follows that she would easily absorb everything that comes her way. This prototype makes Dianne susceptible to acquiring not only viruses and bacteria, but other stresses and strains more easily than most.

On the positive side, her great sensitivity allows her to soak up the aroma of the fragrant red rose and perhaps appreciate its sweetness more than the rest of us. Her subtle impressionability constantly scans and tests her environment. The Two opens itself and accepts the good and the bad life has to offer. Sometimes this number needs a shield to protect it from the harsh world that it encounters. All people like Dianne who are physically subjected to the Two influence could use a little help in taking care of themselves. Fortunately, they usually get that much needed shelter and protection through friends or partnership.

The Philosophy of Three

CREATIVITY

The masculine presence of God and the feminine presence of God are outlined in the numbers One and Two. The male and female energies are the pillars that support all of creation. In analyzing the One and Two, we gain a better understanding and appreciation of these founding principles. Numerology provides insight into opposing male and female perspectives and clearly illustrates how each of them approaches and solves problems. The focus, intentions, and powers of each gender are basically different. In the first five numbers, the masculine principle appears to have a noticeable edge over the feminine. The masculine force takes the lead, but the feminine energy cannot be fully accessed until integrated with the higher numbers.

As you continue with this study, you will find the genders eventually reach a balancing point. The domination of power is then transferred to the feminine before it begins to shift back and forth between the sexes. The flow of energy between the masculine and feminine creates a pumping system and the progressive motion of creation depends on this system. We must realize that the male and female energies can never stay in a state of equilibrium. If they did, the male and female would become mere replicas of one another, and the flow of communication and the exchange of power between the energies would become static. There is a vital tension in the male-female relationship, and perfecting their dance of energy is the focus of our evolution.

Imagine the One and Two standing as perfect columns. Describing them together as the Father-Mother principle is apropos as the One radiates the force of will, and Two emits love in response. An explosion of creativity occurs when these two complementary energies make contact. This event gives birth to the child, represented by the number Three. Like every youngster, the Three needs to run, jump, and play. Three is pure creative expression. Never shy and always noisy, Three frequently laughs aloud simply to release the joy and glory of being alive. Every moment presents a new field of opportunity when seen through the innocent eyes of the child. Three becomes intensely stimulated by all that it encounters, and its boisterous reactions usually arouse everyone who happens to be in the vicinity. The creative Three is never short on ideas, and it is certainly not afraid to express its thoughts. The Three can converse endlessly on a host of topics. When it encounters a difficult task, it does not waste time searching the library for solutions. Three will talk about the problem, and talk about it, and talk about it, until it creates an answer with its voice. Three teaches us by example that creative expression is accessed first and foremost with the voice.

In the opening lines of his gospel, John says: *In the beginning was the Word, and the Word was with God, and the Word was God.* John relays to us in this passage that speech is the essential tool that God uses to birth the creation.

Creativity can be packaged in many different configurations ranging from the artistic, to technical and scientific formats. The inventive nature of Three always gives a new or fresh perspective to life no matter what approach it takes. It loves to laugh, and it scatters its charm, ecstasy, and delight everywhere it goes. World wide, diverse cultures recognize the enchanting quality of the number Three. Universal expressions such as *the third time is the charm,* endorse the concept that three has the ability to activate powerful magic.

Some of our most treasured concepts are packaged in groups of three. When examining the following list of attributes notice that you often see active and passive traits that project a dynamic conclusion.

- Mind, Body, & Spirit
- Father, Mother, Child
- Father, Son, & Holy Ghost
- Thinking, Feeling, Willing
- Mental, Physical, Emotional
- Past, Present, Future
- Morning, Noon, & Night
- Beginning, Middle, & End
- Reading, Writing, & Arithmetic
- Faith, Hope, & Charity

ILLUSTRATIONS

1. 2.

1. When the mother principle and the father principle come together, they create a triangle, the first geometric shape to emerge in this reality.
2. In the modern 3 we see a half circle on the top and the same reflected on the bottom. It appears as if the cosmic egg has cracked in the center and opened wide.

DESCRIPTIVE ADJECTIVES

Expressive, Imaginative, Optimistic, Gregarious, Sociable, Amusing, Artistic, Eager, Joyous, Lovable, Verbal, Youthful, Lively, Animated, Playful, Creative, Original, Comical, Emotional, Friendly, Popular, Spirited, Fertile, Enthusiastic

THREE SPEAKS

- *I have a joke for you.*
- *My friends tell me that I appear much younger than I am.*
- *Would you like to come out and play with me?*

JOB PLACEMENT

Salesman, Actor, Entertainer, Artist, Teacher, Chef, Public Speaker, Recreation Director, Writer, Receptionist, Party Planner, Child Care Worker, Decorator, Demonstrator, Coach, Golf Pro, Caterer, Promotions, Convention Planning, Painter, Minister, Sales

CONTEMPLATE

The Story of Creation/ Day 3

Then God said, "Let the earth bring forth vegetation: every kind of plant that bears seed and every kind of fruit tree on earth that bears fruit with its seed in it."

Note: This passage clearly speaks about the birth of a variety of life and renewing life with seed.

A GLANCE AT THREE

Mythology figure:	Hebe or Juventas, goddess of youth
Zodiac:	Gemini
Tarot:	the Empress
Gender:	Masculine
Mind:	Conscious
Temperament:	Emotional
Color:	Yellow
Element:	Fire
Energy:	Active

The Humanity of Three

The Playful Pup

How difficult could it possibly be to be filled with the creative energy of a child?
What would it be like to be an adult unfettered with worries and responsibilities?

Tom and Alice had been clients for many years before they brought their teenage daughter, Casey, to me for counseling. They were cautious parents who wanted only the best for their daughter. At the time of our first visit, Casey was a freshman in high school. She had begun to neglect her homework and her grades suffered as a result. These parent shared aspirations of their daughter heading off to college one day, so they saved for her education from the day she was born. For years (as many parents typically do) they held a collegiate image of Casey in their minds. Tom and Alice hoped their daughter would some day fulfill *their* dream. Realistically, however, their daughter's spirit was never going to fit into their vision.

After I studied the energetic configurations of Casey's numerology and saw how the number Three strongly influenced her actions, I posed this question to the parents:

"How disappointed will you be if she decides *not* to go to college?"

By asking this question, I was gently preparing them to adjust their plans. Her numerology showed that Casey lacked the motivation to jump through those hoops, to sit and read the books, or to test the waters with a part time job as high school students sometimes do. Casey was not interested in school or work and no amount of convincing or any system of rewards could lure her in that direction. Tom and Alice confessed that a few critical friends and relatives bluntly referred to Casey as a lazy spoiled brat. That diagnosis could not have been further from the truth. Lazy she was not. With all of the Three energy that she was exuding, she could hardly contain herself. It is common for Three to display difficulty in organizing and directing its rich creative thinking, simply because it has such an overabundance of inspiration.

I knew that Casey liked to dance. With that particular talent in mind, I asked her if she had ever considered getting a job as a recreation specialist on a cruise ship or playing a character role at a theme park some day. The thought of professionally promoting fun and games overjoyed Casey. She was not meant to work in the traditional sense, and she knew it! She was not stupid, lazy, or spoiled; rather, she was desperately trying to listen to her higher power. In her own way, she was attempting to follow her divine plan to the best of her ability.

My prescription for this young lady was to get her out of the conventional educational system and into a dance studio for six hours a day. Even after the dancing, Casey would need additional physical activity to satisfy her needs. Ideally, as she danced and exercised, she would be permitted to fully express her emotions. Laughter throughout the day would facilitate an open and honest release of feelings.

Most educators would have had a difficult time accepting Casey's constant output, as this one child generated the creative energy of a whole classroom of children. Sadly, she lacked the patience and maturity it takes to assemble her talents and make use of them in a practical way. Casey needed someone else to direct her creative spirit and give it shape and form. At that time in her life, it would have been advantageous to have had a personal trainer and mentor to give her the attention she required. One-on-one supervision would have helped her to establish boundaries, be productive, and experience a sense of accomplishment. Unfortunately, our society has difficulty accommodating this type of child. Our public educational system isn't geared to nurture high energy, creative children. Instead of developing the talents of someone like Casey, these children are labeled and often drugged.

I last saw Casey when she was a junior in high school. Amazingly, she did not look a day over the age of eleven. By the time Casey reaches mid life, she will probably appear to be twenty five years of age. Typically, people that carry Three energy have an eternally youthful appearance. It is their job, after all, to project the spirit and image of the child.

A MIDDLE-AGED THREE

No matter what age a person may be, it is not easy to manage the high voltage creative power of Three. I have counseled many couples whose marriages suffered due to the influx of the Three vibration in one partner as they entered mid-life.

The case of Linda and Bill presents a classic example. For the past eighteen years this couple has owned a very successful appliance sales and service company. The business demands that either Bill or Linda be in the store to take care of customers on a daily basis. When they started out, Linda focused solely on the office work, but as time passed she also learned to do the repairs. A model of efficiency, this pair shared all of the duties of the business. They achieved success with the dedication and commitment they each contributed to their special partnership. Always honest, stable, and dependable, this balanced duo formulated a no nonsense approach to life, reaching their goals through a willingness to roll up their sleeves and work hard.

The day finally arrived when Linda turned forty two. Her numerology showed that she was pre-programmed to learn the lesson of the number Three at that time. This lesson was scheduled to stay with her for a number of years. Once the Three energy took effect, poor Bill did not know what happened to his steady partner. Overnight it seemed that Linda did not want to work anymore. She would open the store late and forget to write down orders. During this time, Linda suddenly remembered how much she loved animals. She bought a cute little lap dog complete with bows to keep the fur out of its eyes. Linda painted the dog's nails pink, and it proudly accompanied her to the store every day. Although customers loved her little MiMi, Linda was not satisfied. She returned to the pet store and bought a pair of hamsters complete along with an extensive habitat. She also picked up an adorable turtle housed in a state of the art aquarium. Linda thought it would be a great idea to display the pets at the customer service desk in the appliance store. She took excellent care of her little darlings; nevertheless, when I last visited her, I noticed that the appliance store smelled slightly like a pet store.

When the novelty of the menagerie wore off, Linda began to feel bored again. As an escape from the routine, she began chatting on the phone with friends throughout the day. Regrettably, her personal conversations resulted in customer calls being delayed and sometimes business was lost. Linda also started the habit of welcoming the weekend Thursday nights with a friendly hand of poker she arranged with other business owners on the block. Linda signed herself and Bill up on a couple's bowling league because she thought it would be fun. She also started inviting friends and relatives to their home every weekend destroying any chance for peace and quiet on Sunday mornings. Linda knew she was creating upheaval in her marriage, and she thought the problem might be solved by buying a summer home on the shore of a nearby lake. Bill went along with her suggestion and soon they had another house in which to host more parties.

When the Three vibration arrived, Linda began enjoying life more than ever, but it created upheaval for Bill. He was not attuned to the whimsical Three energy, as it was not his lesson to learn. Confusion set in as Bill was left wondering what had happened to his stable life. During counseling, I explained the nature of the number Three and gave them a time table with the number of years that this force would affect Linda's spirit. Neither Linda nor Bill had the power to make the vibration leave. All they could do was accept the lesson intellectually, and learn to cope. For the next ten years the Three energy would continue to be a great test of Bill's love for Linda, and Linda would continue to be riotously full of fun.

The Philosophy of Four

ORDER

The instant the male and female made contact, their conceptual reality changed. This sacred link resulted in the creation of a new paradigm. The dimension in which we presently reside came into being because of this union. The first geometric shape to emerge in this reality was the equilateral triangle, a distinct product of the numbers One and Two. The bonding of One and Two (the male and female) materialized this new life form, and they now have a responsibility to nurture and care for their creation.In human terms, the marriage of the male and female and the birth of their child changed everything. This powerful event established the basis of life as we know it.

Previously, we gave recognition to the father, mother, and child expressed in the numbers One, Two, and Three. With the creation of the family trio now in place, we are compelled to leave the theoretic realm of abstract thinking and head for more solid ground. It's time to address the material needs of the family.

The constraints of the physical world must be acknowledged. The Four vibration requires the new parents to conform to the rules and regulations governing their physical reality. They have material needs to fill such as food, clothing, and shelter. The family also requires a satisfactory location in which to settle. From this place they can continue to work to fulfill their ongoing wants.

All families want to feel secure and protected, and the number Four responds to those demands. A masculine vibration, Four works hard and applies its innate talent as a builder. When this vibration is activated, it painstakingly follows directions and methodically sees to it that all tasks are completed to specifications. The Four takes pride in its role and vigilantly embraces a strong work ethic. Fortitude enables it to plow through boring and distasteful tasks without complaining. Often a moral teacher, Four instructs others on how to do things "the right way." It disapproves of cutting corners and promotes honesty on the job. The good example set by the Four encourages others to do their best to fulfill their duties. Four's tenacious will generates a strong sense of determination. Sometimes, plain stubbornness just won't allow Four to give up on its plans or change its course even in the face of seemingly insurmountable difficulty.

Ever famous for being orderly and consistent, Four takes its place as the great regulator of time. On the cosmic level, it represents the force that keeps track of the movement of the planets around the sun. Beside counting and recording the passing of days and nights, months, and years, Four also sets the basic directions as north, south, east, and west.

Gaining control over the energy and the power of Four furthers the possibility that our creative ideas will take shape. If we open the equilateral triangle of the creative number Three and add one more line, the drawing becomes a square. Expand the drawing with the volume of the three-dimensional world and our linear box becomes a cube. Hence, the cube symbolizes the earth and all of our solid reality.

There is an expression we use when we want to refer to someone who is a solid and upstanding citizen. We call him *"the salt of the earth."* Essentially, we typify him as being shaped like a square to acknowledge that he embodies the stability of the cube, the symbol for salt. When referring to a person as *"the salt of the earth,"* we imply that he can be trusted and relied upon for his honesty, loyalty and accountability. We can always depend upon a person with Four energy.

As a particular note of interest, chemists use the cube to represent salt. Salt, an important building block of all life on the earth, plays a vital role in the construction and regulation of the human body. Our blood is actually a saline solution. Sodium in the body works to prevent water loss, and plays a major role in the transmission of nerve impulses. With these facts in mind, we can suppose that the cube is responsible for the manifestation and structure of our physical body and our planet.

In the context of our study of numbers, the cube, salt, or Four energy are all terms that can be used interchangeably referring to the basic building blocks of life. The solid and dependable Number Four presents the master builder and the force that regulates and maintains our systems. It gives structure to a world of ideas.

Consider the following list of four fold concepts which demonstrate the law and order of our existence.

- **Direction:** North, South, East, and West
- **Elements:** Earth, Fire, Air, and Water
- **Phases of the Moon:** New Moon, First Quarter, Full Moon and Last Quarter
- **Fixed signs of the Zodiac:** Taurus, Leo, Scorpio and Aqarius
- **Evangelists:** Matthew, Mark, Luke and John
- **Seasons:** Spring, Summer, Fall, and Winter
- **Phases of Life:** Childhood, Youth, Adulthood, and Old Age
- **World Ages:** Golden, Silver, Bronze, and Iron Age
- **Human Temperament:** Physical, Mental, Emotional, and Spiritual
- **Financial Statements:** first, second, third, and fourth quarter reports

ILLUSTRATIONS

1. 2. 3 .

1. The square is the geometric shape of 4.

2. The triangle and the T-square combine showing the tools of the builder.

3. Modern 4

DESCRIPTIVE ADJECTIVES

Accurate, Consistent, Dependable, Disciplined, Loyal, Obedient, Productive, Reliable, Honest, Trustworthy, Cautious, Conventional, Factual, Stable, Methodical, Orderly, Conservative, Meticulous, Conforming, Efficient, Dedicated, Law-abiding, Solid, Systematic, Thorough

FOUR SPEAKS

- *Give me the plans and I will build it.*
- *Establishing a routine is important because it allows me to be more productive when the pace of life gets hectic.*
- *I love tradition, and I think it is important that we pass on our customs and values to our children.*

JOB PLACEMENT

Accountant, Maintenance Manager, Construction Worker, Cashier, Quality Control, Systems Analyst, Custodian, Landscape Architect, Customer Service, Laborer, Waitress, Librarian, Truck Driver, Stenographer, Clerk, Bank Teller, Statistician, Inventory Control, Seamstress, Maid, Refuse Removal

CONTEMPLATE

The Story of Creation / Day 4

Then God said: "Let there be lights in the dome of the sky, to separate day from night. Let them mark the fixed times, the days and the years, and serve as luminaries in the dome of the sky, to shed light upon the earth."

Note: The fourth day introduces the concept of organization. God establishes time and provides a system to track its passage in years by watching the stars.

A GLANCE AT FOUR

Mythology Figure:	Atlas, the Titan who supported the world on his shoulders
Zodiac Sign:	Taurus
Tarot:	the Emperor
Gender:	Masculine
Mind:	Logical
Temperament:	Physical
Color:	Purple
Element:	Earth
Energy:	Rhythmic

The Humanity of Four

Salt of the Earth

Juan is a good friend, whom I have been acquainted with for many lifetimes. He currently serves as a hospital administrator. This gentleman has a great deal of Four energy flowing through him, and he serves as the perfect example of how this number can manifest in a person's life. I spoke with him and asked if he could comment on how form, order, and time management fit into his life. This is what he had to say:

Time management is important. My time at work is already scheduled for me, but I like to organize it into blocks. That way I can get more done. I am very busy at my job now. There is so much more required of me these days. Because of this, I don't enjoy the work any longer, but it is something that has to be done. I have always worked very hard, but there was a time that I loved what I was doing. I really felt like a father to the other employees at that time. Now, I don't feel that way, but that is probably because I no longer love what I am doing.

His wife suggested that he tell me about his work week routine. Juan complied: *My routine begins when the alarm rings. I wake up and go into the bathroom. After I shave and shower, I take a shirt from the clean laundry basket downstairs to iron. I set up the ironing board and while the iron is warming I look for the coffee cup, bowl, and utensils I will need for breakfast. I go outside to pick up the newspaper and come back into the house and pour a glass of orange juice. To make the maximum use of my time, I drink the orange juice while I am ironing the shirt. I eat cereal, toast and a banana every day. On Monday, Wednesday, and Friday I eat Total, but on Tuesday and Thursday I will eat some other brand of cereal for variety. On the variety days, I always add wheat germ because other cereals don't have the same nutrients as Total. I like to stick to a formula. Before eating the cereal, I put the toast in the toaster and hit the coffee button. I set up the coffee with the exact amount of coffee beans and water the night before. It takes eleven minutes to make coffee. The coffee and toast are perfect this way.*

There will come a day when Juan retires. When he no longer has to march off to work every morning, his ritual will probably change. At that time Juan will be free to establish a retirement routine.

I asked Juan how he deals with things in his spare time when he doesn't have to stick to a schedule. I wanted to know if he would be uncomfortable deviating from his routine. His unhesitating reply follows:

I am so used to being active. We are really busy here. I would say that I really have no time for that.

The Four energy surrounding Juan dominates and commands organization in every aspect of his life. He does, of course, have *some* leisure time, and his family knows that golf becomes a priority on the weekends. Interestingly, he approaches the game of golf with the same attitude that many people use to tackle a job. Because rituals are a natural part of his mentality, Juan's pre-shot routine has become very polished. He faithfully and consistently adheres to methods his coach recommended years ago. Every shot he takes is given the same serious attention, and he consistently displays a smooth and methodical performance as he moves through his golf swing.

Because he is so intent on keeping track of everything, Juan automatically records the number of strokes he takes on each hole while walking toward his cart. The importance of the record keeping shows in his facial expressions, and when you are in his presence, you can't help but notice him calculating internally. All the important game day information he collects is destined to be transferred to a log book that tracks his scores, weather reports, and course conditions.

A courteous golfer, Juan cautiously calculates the amount of time the people in his foursome take on every hole. He always watches the clock and reminds the group to keep a steady pace. If they are unable to move along in a timely manner, he promptly signals the golfers following them to play through. This inherent urge to stick to the schedule whispers to him constantly. He cannot change the habitual factors about himself anymore than the cat can eliminate its compulsion to chase the mouse.

A SPECIAL NOTE: THE HUMANITY OF FOUR

If you have little or no Four influence in your numerology, Juan's behavior might seem compulsive. As I relayed his story to different audiences, I heard some free spirits unabashedly express their opinions. I found many of these people to be very critical of Juan's behavior and quite emphatic about giving a diagnosis. Many of them said, "Something must be *wrong* with Juan," and "His behavior is not *normal*."

Other people who carry a great deal of Four energy read the same account and found comfort in Juan's methodical behavior. Reading Juan's story made many of these people feel validated. Perhaps now these folks will no longer be shy or apologetic about expressing their very structured Four energy.

Please remember: *Strictly adhering to a schedule does not necessarily indicate a defect of character.*

Consider that the Four energy effectively **structures** our reality. If you are an **unstructured** free spirit who does not like to follow the rules, then bless the Four people, who in effect, do this job for you.

The Philosophy of Five

EXPLORATION

As we saw in the previous chapter, the number Four takes a practical approach to performing its duties. Its mechanical nature constantly works to keep things orderly and to provide us with the necessities of life. Four's energy pulses rhythmically like the beat of the heart, setting a pace that is steady and consistent. It moves us ahead in a systematic fashion. Like the tick of the metronome, it reminds us to place one foot in front of the other as we walk our evolutionary paths. Repetitive motion comforts Four, hence it loves following patterns. The seasons pass sequentially and the orderliness of the universe is accepted without question in this controlled state. The plants and most animal species move along within nature's system without the bother of a questioning ego. We, however, have the seed of *I am* planted within, and that unique aspect of humanity emerges under the guidance of the Five Vibration.

Envision the human figure with the arms and legs outstretched. We see five extensions into the universe (two arms, two legs, and the head). Our head and limbs are our tools to reach outside of ourselves and take us exploring.

The five-pointed star or pentagram can be overlaid perfectly on the human figure and has long been the symbol used to represent evolving man. The top point of the star (the head of the man), indicates the presence and power of the will. From the head, we make our choices and propel ourselves forward into the universe to explore and grow. Note when the pentagram is reversed with the head pointed downward, the same star becomes a symbol of degeneration. For this reason, many esoteric traditions designate the reversed five pointed star as a sign of the devil. The upward pointing star symbolizes progress and connotes good intentions, whereas the downward pointing star indicates the unfolding of negative actions on the path of evil.

Our capacity to choose between good and evil is activated with the Five Vibration. This number not only emphasizes our human individuality, it demonstrates the control we exercise with free will. We live on this earth to experience and learn from our encounters. Five awakens the mind and allows us to expand our awareness so that we can fulfill the unique role assigned to each of us. Five motivates us to step outside the box, go exploring, and find the answers to questions in our own special way. This vibration helps us to realize our capacity to use our senses and to think. The number of exploration and discovery manifests in our physical bodies with 5 fingers on each hand, and 5 toes on each foot. We have 32 teeth in our mouth 3+2=5. We have 5 senses to enrich our exploration as we learn, expand, and gather knowledge.

A feeling of freedom permeates the air when we encounter this vibration. Set routines are challenged and as hidden talents emerge, limitations begin to dissolve. With restriction no longer dominating our behavior, the privilege to move about, and do as we please becomes highlighted. We joyfully engage in the autonomy that the Five affords, however, this new found freedom does come with a price tag. We must also deal with the consequences, both positive and negative, of our actions. Although penalties and rewards exist for every choice we make; human beings derive great satisfaction from the ability to pick and choose their own unique experiences.

Vital to the Five vibration, sexuality activates a broad range of communication and offers the most intimate way we have of expressing ourselves. Using sexuality and sensuality, we are able to explore and learn directly through the experiences of pleasure and pain.

Commonly, you will find a significant amount of movement and change permeating the arena of the Five. Activity becomes the teacher, and each new scenario encountered along the evolutionary path feeds more data into the bank of consciousness. If you stop to take a look at the placement of number five in our string of integers, you will see that it is in the center of all the action.

1 2 3 4 **5** 6 7 8 9

Due to its central placement as the keystone, Five feels the energy of all the other numbers flow through its territory. If we arrange the numbers into a cube, the Five continues to be centrally located. Just as five receives something from every other number, the energy of five is transmitted to all the other numbers.

1	4	7		1	2	3
2	**5**	8		4	**5**	6
3	6	9		7	8	9

Up-to-the-minute information is constantly being processed by the Five, influencing all decision making. This continual collection of data generates new knowledge. Because of its central position, the five has been aptly titled *"the great communicator."*

ILLUSTRATIONS

1.

2.

3.

4.

1. the figure of man
2. the five pointed star
3. the symbol of Jupiter (the fifth planet)
4. modern 5

DESCRIPTIVE ADJECTIVES

Energetic, Resourceful, Active, Adventurous, Changeable, Free-Spirited, Productive, Progressive, Spontaneous, Quick, Inquisitive, Daring, Dynamic, Exciting, Curious, Lively, Gregarious, Demonstrative, Eager, Clever, Imaginative, Sexual, Sensual

FIVE SPEAKS

- *There is nothing better than the thrill of a good adventure.*
- *If I'm not multi-tasking, I get bored.*
- *Can I have a taste of that?*

JOB PLACEMENT

Travel Agent, Sales Representative, Receptionist, Advertising Agent, Buyer, Editor, Trouble Shooter, Personnel Officer, News Reporter, Transportation, Interpreter, Teacher, Linguist, Broadcaster, Speaker, Publisher, Middleman, Marketing, Politician, Pilot

CONTEMPLATE

The Story of Creation / Day 5

Then God said, "Let the water teem with an abundance of living creatures, and on the earth let birds fly beneath the dome of the sky." And God blessed them saying "Be fertile, multiply, and fill the water of the seas: and let the birds multiply on the earth."

Note: During the fifth day of creation, God places a variety of creatures on the Earth and blesses them with the ability to reproduce. As the great communicator, and the number of physicality and sexuality, Five demonstrates its abilities perfectly in this passage. Five utilizes the senses to explore, communicate, and further evolution.

A GLANCE AT FIVE

Mythology Figure:	Mentor, tutor and advisor of Odysseus
Zodiac Sign:	Aires
Tarot:	the Hierophant
Gender:	Masculine
Mind:	Reasoning
Temperament:	Mental & Physical
Color:	Orange
Energy:	Explosive

The Humanity of Five

A Jack of All Trades

My family has the very good fortune to be acquainted with a wonderful Texan named Don. He happens to be one of the most amazing individuals in my cast of characters. When I was searching for a real life description of the number Five, my thoughts went immediately to him. From what I already knew about Don's life, he seemed to be the perfect example of the humanity of the number Five. The only problem was that I had never professionally studied Don's numbers. So I called his wife, Paula, to chat and get the needed information to calculate his chart. This was a wonderful opportunity for me to test the validity of numerology. The perspective I had of Don made me feel confident that I was going to find a chart saturated with Fives. It was important to verify that the number Five was significantly represented in his life.

As I carried out my little test, my expectations were more than satisfied. Reviewing his numerology, I was quite overwhelmed to find so many Fives in key positions. Seeing all of his Fives on paper helped me to better understand Don and the reasons for the decisive choices he had made over the years. The chart also allowed me to accurately project how these influences are shaping his probable future.

During our college years, Paula introduced me to her then boyfriend, Don. His tall, handsome, and kind radiance made him appear to be a modern day superman. He was evidently a strong physical type, as he ran twelve miles a day. (It was six miles to visit Paula; then another six miles for his return home.) With a busy schedule playing basketball for his division one school, he still managed to find time to work as a lifeguard. Always sensually involved in the art of living, Don needed to experience life with every fiber of his being. This Renaissance man was certainly not lacking when it came to mental prowess. He loved puzzles and solved mathematical problems for recreation. His abilities lead him to earn a Masters Degree in Statistical Analysis. For Don, then and now, any work that involves calculus is a pleasure. In his thinking, calculus provides both a mental challenge and laid-back entertainment combined in one package.

Paula told me when their sons were in high school, Don always looked forward to helping them with their homework. She said he would wait for the boys to come home from school specifically so he could do math homework with them. Although the they enjoyed this routine for awhile, eventually his sons did not want or need his help because they were just as brilliant as their father. This was a very depressing time for Don because his special opportunity to share his love of mathematics with the boys suddenly ended. Situations change as children grow up and parents adapt. Today, Don continues to espouse calculus to all who will listen with his motto, *"Calculus makes sense!"* Many students have never used their advanced math after leaving college, but Don continues to incorporate this knowledge into his daily life.

Currently Don works for a company that sells a product that helps oil flow more easily through the pipeline system. He explains that when they pump oil out of the ground it is not necessarily in a liquid state. It often surfaces as tar or a thick sludge. His job is to figure out how much of the friction reducing product his company manufactures is needed and at what pressure it must be injected into the pipeline to efficiently move the crude along. Don's knowledge, experience, and skill have enabled him to become an expert in the science of hydraulics, and he is particularly well suited to his job.

Obviously, his profession requires a strong mental type and Don has proven himself in that arena. Since Don embodies an abundance of Five energy, he is also extremely physical. This aspect makes him refuse to stay put in the home office. Don reasons that in order to insure the accuracy of the field reports, it is imperative that he go to the job sites to collect his own data. His insatiable need to be active and explore makes business travel a welcome part of his life. Maintaining his schedule can be quite hectic but the constant travel doesn't leave him feeling drained as it often does to other employees.

His domestic travel routinely takes him to Alaska, Washington, Oklahoma, New Mexico, and California. In the past several years he has also spent countless hours working in Poland, Russia, Siberia, Ukraine and Azerbaijan. In addition he has made some minor trips to South Korea, Japan, Chili, Brazil, and Argentina.

The high adventure of this number Five man reads like something out of a James Bond novel. His exotic travel destinations and the rigorous demands his work puts on his mind and body are exceptional. The position requires that he be dynamic, resourceful, inquisitive, and quick thinking. All of these qualities match perfectly with Don's five nature.

Company policy mandates that Don undergo periodic physical training exercises. These training sessions serve to test the physical strength and response times of each employee. One of the drills necessitates that Don be blindfolded and put in a smoke filled room. The objective of this test is to find the exit before he expires. Another exercise Don related simulated his helicopter malfunctioning and going down in the North Sea. To accomplish this, he is tossed upside down in a tank of ice water while strapped to a helicopter seat. Again the objective is to survive. In this case, escaping the seat and swimming to the surface becomes the goal. Those college years of working as a lifeguard have continued to pay off. The ingenious, resourceful Five energy never ceases to amaze.

Don is unmistakably a classic number Five, a jack of all trades and a master of them all. When at home relaxing on the couch, he will often watch several channels on the TV while surfing the internet and discussing business on the phone. True to form, his Five nature thrives by keeping all the senses stimulated.

The Philosophy of Six

LOVE

Taking a moment to digest the numerous experiences gathered under the guidance of the Five, we find ourselves on a bridge. This passage leads us on to more complex undertakings. At the junction of Five, we leave the lower order numbers behind (1-2-3-4) and ready ourselves to explore the complexity of the higher numbers (6-7-8-9). We can appreciate how the father-mother-child triangle was formed and understand how family obligations were methodically taken care of under the Four vibration. Although logistical needs were met, man was not satisfied. He felt an innate desire to explore and discover more. The Five energy encouraged him to follow his instincts and compelled him to extend his world through communication, exploration, and experimentation. A variety of passionate cravings propelled man to travel and learn more about his own nature, relationships, and the world in which he lived. A myriad of possibilities were tested as man, enlivened under the power and influence of the Five, began trusting the information he gathered with his senses. Direct experience became the catalyst for growth. Under the tutelage of Five, man came to realize his ability to draw valid conclusions by intellectually processing his discoveries. At this stage, his guiding force became his own thoughts. Once he was able to acknowledge the power of his own free will, man was finally able to reach for the comfort of his heart.

As we continue our journey in this sequential study of numbers, Six teaches to us to appreciate the power of the heart and its ability to facilitate our growth through love. The acceptance of love is a huge step on the evolutionary ladder. We may speak of it as a process, yet it seems to take place instantaneously.

A simple analogy will help to clarify this seeming paradox inherent in the sixth step of our evolutionary development. Imagine the action of an archer pulling the string back on his bow. In the second it takes to release the bow string, man is able to turn inward and reunite with his heart. An instantaneous sequence of events plays out that allows him to gain a renewed sense of security and satisfaction. All of his previous searching draws to a climax and propels him directly back to the target. (Recall, the target in our numerical studies is the self.) In the blink of an eye, man realizes that the heart and mind are united. Suddenly, he becomes comfortable and confident with his discovery that he contains both the male and female energies within his being. He can finally appreciate the intimate working relationship between these two aspects. At last man recognizes the capability he possesses, and with his new found awareness, he commands more of his world.

An understanding of the meaning of the number Six can also be illustrated graphically. Six can be divided into two equal parts or drawn as equilateral triangles in sacred geometry.

The interlacing of the two triangles creates a six pointed star which directs our attention to two different pathways. The symbol is in effect a mirror with the tip of one triangle aiming up to the heavens and the tip of the opposite triangle pointing down toward the earth. From this magnificent star, humanity has formulated the expression, AS ABOVE SO BELOW. The Six pointed star speaks of the fact that we are a replication of our divine inheritance. It perfectly illustrating the seamless bond humanity has with its celestial parents.

The balance of the male and female elements (the balance of the head and the heart) can also be surmised in this design. In Eastern theologies, the triangle pointing upward represents the principle of fire and thoughts. The downward pointing triangle symbolizes the element of water and the emotions. When we interlace the two we construct an image that has been labeled in different cultures as the Star of David, Solomon's Seal, or the Philosopher's Diamond.

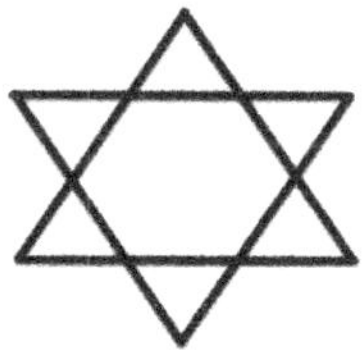

The six-pointed star unequivocally signifies our inherent ability to balance the spiritual and physical worlds. In striving to attain this equilibrium, humanity must come to terms with both the eternal and the ephemeral nature of creation. Working independently, the logical mind is not capable of embracing and understanding ***All That Is***; however, when the heart operates in conjunction with the mind, the undertaking becomes plausible.

The logical mind operates by shifting around blocks of knowledge it already possesses. The understanding supplied by the heart enables thinking to expand in ways that the mechanical mind could not possibly conceive. There are numerous historical examples we could site where scientific thinking screamed ***"Impossible!"*** even as others worked to achieve seemingly unattainable goals. Walking on the moon and instant worldwide communications were once considered ***impossible***. Because the heart does not depend on reason and logic, the concept of ***impossible*** doesn't resonate with it. The heart blesses and accepts what the mind has to offer, and it also nudges the mind to move beyond its accepted boundaries. An attitude of appreciation and mutual respect between the mind and heart motivates us to love and understand ***All That Is***. As love emerges in our thinking, humanity is elevated in status.

When our emotional world merges with our mental processes, our perspective shifts substantially and changes the approach we take to fulfill our needs. At one time man built a house to have shelter from the elements. Now, utilizing the Six energy, he is able to entertain an expanded view of the purpose of that house. The shelter evolves from a safe haven where family members can gather, to an ideal place where they share their dreams, heal, and gain strength to face life's challenges. With the presence of the Six vibration, the ***house*** is elevated to the status of a ***home***. Man's expansion of awareness helps him to realize that, besides providing shelter, his dwelling has the power to nurture and comfort his soul. We witness this transformation of consciousness when we watch someone mowing the lawn, planting flowers, painting, and otherwise upgrading his property. Whether that person personally labors on the tasks or pays another to do the work is not important. The shift of energy and the intention to make the place a work of art is driven by the Six energy. The creation of a loving environment becomes possible when man accepts and balances his male and female qualities. The emergence of beauty into the atmosphere takes place effortlessly once these aspects are fully integrated within a person.

The balanced and elevated consciousness of the Number Six is reflected in the union of the mother and father principles. Performing the role of a dutiful parent, Six willingly takes on great responsibility. Six provides warmth and care and serves as the healer, the teacher, and the one who nourishes others. It also expands our reality by introducing the concepts of grace, beauty, love, and romance into our world. Under the guidance of this vibration, man can finally accept his divine inheritance. The expansion of awareness generated by Six becomes the foundation upon which greater truth will continue to unfold. The knowledge that we are holy people is ready to emerge.

ILLUSTRATIONS

1.  2.  3. 6

1. The six pointed star
2. Man standing beside the heart
3. Modern six –The one (male) and the heart (female) are combined.

DESCRIPTIVE ADJECTIVES

Responsible, Romantic, Artistic, Domestic, Protective, Healing, Affectionate, Moral, Virtuous, Benevolent, Compassionate, Caring, Emotional, Hospitable, Resolute, Charitable, Conscientious, Demonstrative, Honorable, Hospitable, Kind, Tranquil

SIX SPEAKS

- *I love you.*
- *I'm sure we will both feel better if we sit down, have a cup of tea, and talk.*
- *It feels so good to be home.*

JOB PLACEMENT

Mother, Father, Counselor, Minister, Interior Decorator, Hotel Management, Doctor, Nurse, Gardener, Teacher, Florist, Artist, Chef, Customer Service, Host/Hostess, Social Worker, Healer, Baker, Real Estate Agent, Farmer, Caterer, Waiter, Photographer,

CONTEMPLATE

The Story of Creation / Day 6

God created man in his image, in the divine image he created him; male and female he created them.

Note: The fact that God created man in his image (male and female) extols the principle of, *as above, so below.*

A GLANCE AT SIX

Mythology Figures:	Venus, goddess of love and beauty
	Juno, goddess of marriage
Zodiac Sign:	Pisces
Tarot:	the lovers
Gender:	masculine-feminine balance
Mind:	heartfelt
Temperament:	emotional
Color:	blue-green
Energy:	serene

The Humanity of Six

Mother and Father

The vocation of parenthood manifests under the care and direction of the Six vibration. The Six commands high expectations of each and every person that steps into this role. Once a person gains the title of Father or Mother, he or she automatically becomes a model for his or her children to follow. The moralistic Six preaches to those who are engaged with its energy that vigilance and good judgment are of utmost importance for all Mothers and Fathers.

An examination of the spectrum of Six energy plainly reveals the significant responsibilities all new parents have to fulfill. The position often requires them to serve as cooks, housekeepers, caregivers, gardeners, coaches, teachers, providers, hosts or hostesses, nurses, doctors, and mentors to others in their family circle. Although the demands on parents are high, I find that most Six types tend to develop a specialty. The following examples present two couples who clearly express some of the many ways in which Six energy can manifest through different personalities. As you read these accounts, keep in mind that each one of the individuals in these partnerships is exceptionally dominated with Six energy.

Fred & Florence

Fred and Florence have been married for fifty nine years. One of the many nurturing talents Florence excels at is an ability to communicate with babies. This sweet lady can switch into a special interactive baby mode at a moments notice. Florence can cluck and coo for hours while going through endless patterns of facial expressions. With her fluency in baby language, she could soothe the crankiest infant. Throughout her life, Florence has always related well to very young children and was in her element when working with babies, toddlers, and preschoolers. As the children grow up, she continues to make an effort to give them what they need; however, her joy was always engaging with babies.

The multi-talented Florence has always been quite the baker, and the delightful aroma of homemade cookies fresh from the oven often permeates her kitchen. She likes to be prepared with cakes, cookies, coffee, and tea for the friends and neighbors who might drop by. For many years her magnetism for company proved to be quite overpowering, and she always has someone sitting and chatting at her kitchen table. As she has aged, her children and grandchildren have grown up and many of her old neighbors have passed away, but her generous hospitality and compassionate energy continues to need an outlet. Discontent unless caring for someone, Florence noticed one of her overworked neighbors was unable to devote enough attention to his dog. Florence began to feed the dog and bring him into her house during spells of bad weather. Today, the dog occasionally goes to visit his owner, but he gladly stays with Florence who willingly cares for his every need.

An interesting comparison to Florence's Six energy, her husband Fred expresses the same vibration in a radically different manner. When you are with him, the typical warm and caring Six energy does not jump out and make itself apparent. Being rather high strung, fussing over children never ranked high on Fred's priority list. Fred happens to be one of those people who get nervous around children. He simply does not have the patience to deal with the creative mess kids can generate. Nevertheless, Fred and Florence had four children together. Early in their marriage, Fred passed along all of the duties of raising their children to his wife. Florence, comfortable playing the role of super mom, did not mind doubling up on the parental responsibilities. At the time, this arrangement worked for them. Because he traveled extensively with his job, Fred had limited time at home. This circumstance, however, did not prevent him from enforcing a strict rule forbidding the children to keep any pets. He always seemed to be stretched to the limit with the stress of his job and the antics of four growing children, so he was not about to put out the welcome mat for any cats or dogs.

My quick sketch of Fred up to this point noticeably lacks many of the warm personal qualities usually listed as characteristics of the Six vibration. With so many sixes dominating his chart it does seem odd that Fred didn't express more warmth toward his family. *Why was this so? What happened to Fred's Six energy?*

Carefully examining his chart and the outlets he chose for his talents provides the necessary pieces to this numerology puzzle. Fred never lacked Six energy. In fact, he has much more of this force than any one person can be expected to handle. Although he did not appear to be the model father to his personal family, he certainly played out that role to the greater community. In addition to holding a professional position as an engineer, Fred also managed to become a full time politician for a period of thirty five years. He served as Chairman of the Board of Supervisors in his local community and performed the duties of Chairman for the Council of Governments for the county for many years. Fred had an enormous share of responsibilities with these two positions. His life was saturated with meetings, yet he was driven to accept the position of President of the local volunteer fire company and maintained that rank for fifteen years. Maintaining these high profile positions enabled him to convey his paternal energy to everyone in the community. Fred took calls every night from residents who had problems with their roads, storm sewers, etc. Under his watch, all of the unimproved roads in the township were paved, and a spacious office building, town hall, and a state of the art water treatment plant were constructed. Fred's personal responsibility in keeping the township fiscally sound earned him the reputation of being a loving and dedicated "Father" to his township. Fred carried such huge reserves of fatherly energy that he found it impossible to channel it into one family unit. Responding to the needs of the greater community, he found his true calling. From the perspective of numerology, it is common to find politicians like Fred who have a saturation of Six energy in their charts.

As a special note, I would like to say that Fred's devotional Six energy could have easily been applied to a profession such as doctor, teacher, or minister. We frequently see dedicated people in these positions becoming consumed with the responsibilities of their jobs. Caring for the community family presents a truly difficult and noble calling for any father or mother.

Kevin & Kathy

Kevin and Kathy live in a beautiful home in the mountains overlooking a scenic pastoral valley in rural Idaho. At the peak of midlife, this pair suddenly became deluged with Six energy. Their story describes how the influx of this energy has affected them both.

Although this couple loves spending their days at home on their ranch, come fall, Kathy travels to town every day to teach school. Kathy evaluates and instructs very young children with learning disabilities. The many levels of her job provide her with ample opportunity to radiate that special love and caring Six energy to desperately needy students and their families. Although working with handicapped children expends a tremendous amount of Six energy, Kathy never seems to exhaust herself. After work, she hurries home to tend to her flower gardens, and then pours her heart into quilting and her ongoing home improvement projects.

Kevin manifests the Six energy much like Kathy does when it comes to caring for their home and surrounding property. While she busies herself making decisions about new shades of paint and updating the décor of the house, Kevin can usually be found outside repairing the barn or maintaining the ranch's lengthy driveway. Kevin is much like his wife Kathy in that his caring nature cannot be contained.

In the winter, after he plows his own driveway, he travels up the valley and plows for many of his neighbors. People have come to rely on his generous spirit to rescue them when they are snow bound. Particularly heavy snowfalls in previous winters have left neighbors trapped in their houses. Without Kevin's help, there are a few older people who would have remained confined for a week or two. Kevin's intuitive sense usually alerts him when his assistance is urgently needed, and he makes himself available at those times. He regularly lends a hand to help neighbors who are having problems with their wells. His talents frequently lead him to assist in repairing fences with other ranchers in the area. Kevin's presence offers comfort and support in a variety of circumstances whether that means staying awake all night watching over mares about to give birth, or helping load hay into a barn before it rains.

Both Kevin and Kathy constantly radiate an abundance of loving, caring and sharing Six energy. With all the responsibilities this couple assumes, they still manage to reserve a couple of weeks every August to give something back to their community of ranchers. Together they have created a tradition by hosting an annual horse polo event. The Six energy channels perfectly into the role of host or hostess, motivating Kevin and Kathy to extend an invitation to many families in the area so that everyone can come together for a day of fun and celebration. Their long mountain driveway is always packed with horse trailers every year at polo time. Kevin and Kathy enjoy taking care of the crowd by putting on a big barbeque for those who attend this late summer get-together. The people in the area appreciate their kindness, and many travel quite a distance to attend this affair.

Kevin and Kathy both have a powerful need to be the giver, and oddly enough, their great generosity has caused some conflict between the two of them. Each partner in this marriage constantly strives to do more to care for the other. They both covet the position of caregiver, and neither one is interested in being the receiver. Kevin in particular, wants to do everything for Kathy and constantly feels he cannot do enough. He badgers her at times asking: *Why can't you let me do that for you?* He wants to do all the laundry, take charge of the shopping, and prepare meals. Although she works during the school year, Kathy's dominant Six energy makes her feel a need to participate in these chores. Perplexed with his need to do everything, she has expressed a sense of feeling useless if she does not contribute to the household by helping with the daily routine. She related this to me, *Sometimes, I feel like a visitor with no purpose.*

The number Six attunes both Kevin and Kathy to a parental role and neither partner feels the need to be "cared for". This strong appeal does not indicate an imbalance of character, as Sixes find their balance by connecting with those people who are in need of attention. After reviewing the numerology codes for this couple, I had to forewarn Kathy that her husband's yearning to nurture others is going to get worse! One day, her Six energy will wane, but at that time, Kevin's Six energy will double. It is important that Kathy learn now to adjust to a more receptive role. For several years, I have been encouraging her to sit back and accept the love and care that Kevin wants to give. Assuming the role of queen has not been easy for Kathy. I cannot help myself from thinking, *"How many women would love to have her problem?"*

Kevin and Kathy have two grown children and five young grandchildren, and they have been willing and able to help both of their adult children with their households. I have watched them joyfully spend hours painting, remodeling, and refinishing furniture for their daughter and son-in-law. Of course, they also splurge much of their nurturing energies on the grandchildren, whom they absolutely love to baby sit. These lucky children get plenty of positive attention from these wonderful grandparents. When they are together, Kevin and Kathy laugh and play with the young ones until it is time to kiss them goodnight. When they must finally say goodbye, these grandparents are already making plans for their next visit.

The Philosophy of Seven

SPIRITUALITY

In the beginning of our numerology-creation story, the male force led the way. Masculine energy unmistakably empowered humanity and provided us with our basic needs. Finding the feminine force within and engaging its power then became the underlying drive of the whole evolutionary process. Number Six brought our attention to the great equilibrium of the male and female forces of God. The onset of the Six vibration meant that the masculine and feminine aspects of God were being accepted and balanced within our being. With the work of creation finally completed on the sixth day, love took charge of our continuing evolution. Reaching this momentous stage in our development caused us to pause in acknowledgement. This pinnacle of achievement allowed many advanced attributes to emerge within humanity. The incorporation of the Six vibration enabled compassion, kindness, grace, beauty, affection, love, and devotion to unfolded in man's heart. Processing this new emotional reality was an overwhelming experience; therefore, The Creator allocated a period of time for man to integrate these blessings.

Appreciating these awe-inspiring gifts is the Seventh step on our evolutionary ascent. Although the work of creation was completed in six days, the peace and tranquility following the cessation of activity functions as an essential part of the creation. The number Seven compels man to contemplate his evolutionary journey by pausing for a period of introspection. In the past, many subtle nuances had enabled him to grow and expand his consciousness. Now he must integrate his expansion before he moves forward. An assessment of his progress must take place before he continues, so in this interim he takes time to appreciate how much he has grown. The Seven vibration encourages him to intellectualize his experiences. Man's reflection brings him to the quiet realization that he is *in* the world but not *of* the world. In the quiet of his meditation, he begins to appreciate himself and all that he has learned. Fueled by this new found respect, man expresses praise and gratitude to God for his awe inspiring creation. The Seven energy enables him to pause in a communion of love with the Divine

Seven fosters physical passivity, and encourages man to study and spend time alone for the purpose of introspection. The university campus often becomes a refuge for this individual. He craves the quiet of the library and the sanctity of the chapel. The Seven vibration returns him to an internal place of stillness which he now only vaguely remembers experiencing before he embarked on this evolutionary journey. Through prayer and meditation man gains a deeper appreciation of his spiritual nature. Under the Seven vibration, man fully understands that the manifestation of all life springs from the God within. With his attention on the higher aspects of being, he quietly studies and interacts with all the knowledge he has collected. Living under Seven's spiritual influence, man is encouraged to acknowledge his own holiness. Man's realization of his own divinity brings a sense of conclusion to the creation.

Although not a physically demanding number, the rest Seven provides continues to fuel the ongoing creative processes. If we do not honor the stillness, we cannot further the creation. A simple analogy can be drawn to our basic requirement of sleep. If we do not afford ourselves the necessary rest, our faculties become dull and we do not function to our fullest capacity. The deeper the rest we attain, the better we perform. Scientific research verifies the importance of deep sleep to our sense of well-being. Experiments have shown that meditative rest multiplies our creativity. When we immerse ourselves into stillness, our thinking processes become significantly enhanced.

Consider the implications of expanding our requirement for rest to include the needs of society as a whole. It only makes sense that a communal day of rest would play an important role in balancing any society where groups of minds work together. Because of economic concerns, our society has gotten away from the idea of a common day of rest. Businesses are expected to be functioning twenty-four hours a day, seven days a week. When the group mind has no designated time for group rest, the society becomes weaker and less able to further creation. Without a common day of rest civilization slowly deteriorates.

There is a cyclical nature to the number Seven. We experience this in the rhythm of our 7 days each week, repeating for 52 weeks in a year (5+2=7). We read about the cycles of Seven In the bible. Leviticus Chapter 25 makes a reference to the sabbatical year and the jubilee year:

Verse 3 states:

For six years you may sow your field, and for six years prune your vineyard, gathering in their produce. But during the seventh year, the land shall have a complete rest, a Sabbath for the Lord, when you may neither sow your field nor prune your vineyard.

Verse 8 speaks of the Jubilee Year:

Seven weeks of seven years shall you count—seven times seven years—so that the seven cycles amount to forty-nine years. This fiftieth year you shall make sacred by proclaiming liberty in the land for all its inhabitants.

Completeness and holiness permeates our beings when we consciously recognize the number Seven. Seven is repeated in nature, in time, and in man's understanding of himself and his world. A list of important groupings of seven follows:

7 NATURAL SCIENCES

3 of the creative mind (grammar, logic, and oratory)
4 of the structural world (arithmetic,music, geometry, and astronomy)

7 SACRAMENTS

3 for the spirit (baptism, confirmation,and communion)
4 for the physical (penance, orders, matrimony and extreme unction)

7 DEADLY SINS

3 of the mind (pride, wrath, and envy)

4 of the body (lust, gluttony, avarice, and sloth)

7 VIRTUES

3 supernatural (faith, hope, and charity)

4 cardinal (prudence, justice, temperance, and fortitude)

ILLUSTRATIONS

1. 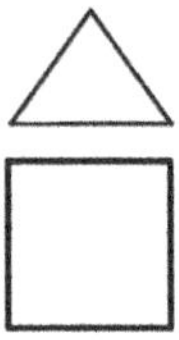

2. 

1. The triangle or the creative forces of three, hovers above the cube which represents the earthly elements.

2. Modern seven, the activity of the vertical bar (male) is topped with the horizontal bar (female) indicating rest.

DESCRIPTIVE ADJECTIVES

Analytical, Scholarly, Spiritual, Introspective, Meditative, Philosophical, Wise, Calm, Intuitive, Observant, Discriminating, Peaceful, Deep, Mature, Conscientious, Aloof, Critical, Peaceful, Logical, Mental, Reflective, Detached, Inquisitive, Scientific

SEVEN SPEAKS

- *Om.*
- *It is mind boggling to imagine that our world is like a speck of sand in relation to all of the teaming activity we perceive in our galaxy.*
- *I find a periodic retreat to the country to nourish my body, mind and soul to be an absolute necessity.*

JOB PLACEMENT

Scientist, Rabbi, Librarian, Monk, Nun, Numerologist, Tutor, Engineer, Researcher, Professional Student, Astronomer, Astrologer, Psychiatrist, Pharmacist, Dean, Coroner, Technical Writer, Priest, Attorney, Mathematician, Anthropologist, Philosopher

CONTEMPLATE

The Story of Creation / Day 7

God blessed the seventh day and made it holy, because on it he rested from all the work he had done in creation.

Note: We see that rest and work are both vital parts of the creation. Work supplies the motivation to rest and vice versa.

A GLANCE AT SEVEN

Mythology figure:	Kronos, god of the harvest
Zodiac Sign	Aquarius
Tarot:	the charioteer
Gender:	masculine-feminine balance
Mind:	logical & intuitive
Temperament:	spiritual
Color:	indigo
Energy:	at rest

The Humanity of Seven

The Thinker

Although I have a very broad client base, it was not easy for me to choose someone whom I thought was a good representative of the number Seven. This highly refined vibration is a bit uncommon, and very few people fully embrace its energy. Seven can create quite a challenge for the person who attempts to find peace with this conscientious, spiritual vibration, and at the same time strives to cope with the demands of our fast paced world.

I have advised a person named Harrison for an extended period of time. The number Seven shows very prominently in Harrison's name and birth date making him very private, refined, and scholarly by nature. In gathering information for this book, I interviewed many people whose lives strongly illustrated one or more numbers. I encouraged these people to give their perspectives on how they felt a number impacted their lives. Everyone I spoke with was very open and immediately shared very personal information concerning his or her character. Harrison was the only one who delayed his response, telling me he wanted to put thought into the questions I posed. He said he knew how important this work was, and wanted to give it the depth of attention it deserved. His approach to the interview distinctly reveals his introspective nature.

Harrison's life growing up could be viewed as one of privilege, with the advantage of private schools, music lessons, and coaching. He had a wonderful loving family with five siblings to challenge and defend him. There were always friends eager to go out with him. This was a life that many would envy, but a life in which Harrison was not fully engrossed. His intuitive spirit has kept him on a lifelong quest for higher knowledge and the peace that only the meditative state can deliver. Harrison was never one to enjoy crowds, and going to parties and social events usually took him out of his comfort zone. To this day, good friends and relatives persist in labeling him a "hermit."

Harrison had this to say regarding the title others have bestowed upon him: *I do like to spend time alone. I think a lot. In fact, I think all the time, and I get a kick out of what filters through my mind. The piano can be a huge escape for me. I love to sit and play. Sometimes I think about going back to music school. I know I would really enjoy taking more instruction. I don't know if I am a perfectionist, but I always want to do better.*

Seven influenced people view education as a necessity. Unsurprisingly, Harrison repeatedly wove this theme into our conversation. If a topic interested him, he would always refer that subject matter back to lessons he took at one time or another. Sometimes he would relate how he loved to study that subject.

As our conversation rambled on, he thought it was important to convey to me how he finds peace in tending his flower garden. *I enjoy the solitude. I love walking the backyard path to the lake at this time of year. It gives me a chance to inspect the trees and to appreciate the seasonal changes on the property.*

He then reminded me that he did take classes and became a Master Gardener. Harrison said that his family was always amazed that he took such an interest in plants. As a child he would sit for hours studying huge volumes illustrating the variety within the plant kingdom. Harrison taught himself the Latin botanical names of all of the local species.

I am sure his knowledge is extensive; however, I have never found any telling evidence of actual physical labor about him. His hands are delicate with no signs of being calloused from tilling the soil. I can't imagine that he would ever tolerate dirt under his fingernails. I am focusing on this issue to emphasize that gardening or anything else that the Seven person gets involved with tends to be more conceptual than physical.

Sevens spend so much time thinking that their reality often exists only on the mental plane. This can be problematic when it comes to building relationships. They think about interaction but often fail to step forward and meet the other person on common ground. Seven types are inclined to be better at wooing their true love with letters than actually arranging a physical encounter. They are often tagged "loners" because they live inside their own heads. Young children who carry the vibration tend to be introverted and the Seven influence often causes them to create *imaginary* friends.

As we talked more about this need for solitude, Harrison explained he prefers working by himself or with small groups. In his professional life, he always tries to avoid participating in big meetings, finding these gatherings to be unproductive. Harrison does not like to waste his time, and is very selective regarding those he schedules into his day. I am pleased he recognizes the importance of numerology and decided to participate in my research.

The Philosophy of Eight

AUTHORITY

Progressing through our study of the numbers, you can observe how the characteristics presented in each number unfold naturally as a result of previous influences. You are acquiring a rich vocabulary in the language of numbers as new concepts that logically build upon previous experiences are introduced. As you might expect, the Eight incorporates a bit of all of the preceding numbers. Prior to approaching Eight, Seven gave reason to pause. As you recall, the Seven vibration holds the experiences of rest, reflection, meditation, and appreciation for the Creation. Although a clear sense of completion resonates with Seven, the two remaining numbers afford us the opportunity to continue to grow and contribute as co-creators.

All higher order numbers (Six through Nine) have the male and female forces equally represented. Each gender definitely displays individual strengths and talents. When the male and female form a united front, they produce and command more power than they were able to generate working independently of one another. The Numbers Six, Seven, Eight, and Nine manifest a higher frequency because of the attunement of the male and female energies. As we examine the effects of these higher numbers further, we can see human consciousness evolving to a more advanced state. The balance of the male and female forces empowers humanity to accomplish more than we could ever hope to achieve had they not stabilized. In the safety of this secure environment, the female influence effectively moves to the foreground to offer her talents to society. The numbers Eight and Nine provide both the male and female a platform to influence and lead others.

Eight embraces and orchestrates all of the power and ideology of the previous energies. The numbers leading up to this point have advantageously prepared the stage, and now all the codes are ready to be used expressly for Eight's purposes. Master Eight can now make its grand entrance and put the principles of creation into motion. This maestro is about to tap the baton and conduct the orchestra as it plays out the symphony of life. Eight accepts the position of controller of all other numbers, and as the overseer, it shoulders great responsibility. It must direct all of the other vibrations, telling them when to interject their powers and just how much force to use. This administrative authority is expected to be knowledgeable concerning the individuals it directs. Caring for the livelihood and welfare of those who work under its control presents a huge task for the number Eight. Eight must possess both physical and mental stamina along with raw courage to face the challenges presented to it. Dependability and trustworthiness emerge as principal assets of this "take charge" number.

When Eight heads a project, body, mind, and spirit are

completely involved. It is not afraid to roll up its sleeves and dive in to demonstrate how a job should be done. It's fearless when it comes to tackling large-scale projects and dealing with a staff of employees. Eight projects high expectations and gets results. Inherently a good money manager, it follows the rule that one must spend money in order to make money. Eight jumps into the role of the consumer with gusto. When Eight goes shopping, it opts for quality and durability in its selections, displaying an inclination to select the finest fabrics and furniture that is built to last. Eight retains high standards because it embodies quality; therefore, it expects to find the same in the surrounding world.

Outside of material concerns, we find that Eight approaches other aspects of life with an all or nothing attitude. When the heart becomes involved, the bond of friendship is taken seriously, as Eight refuses to get involved in ambiguous relationships. It always wants to be sure where it stands so it can project the probable outcome of an alliance. We can predict with a high degree of accuracy that Eight will be uncomfortable casually living with a partner without the benefit of marriage. Most male and female representatives of Eight prefer to go through a traditional courtship followed by the ritual of marriage. Unquestioningly, Eight accepts children as a part of the marriage package.

The Eight portrays its status as an icon of power wherever it manifests. To further illustrate Eight's unbounded potential, consider the symbolic representation of this number. Viewing the Eight horizontally, we are presented with the symbol of eternity, the lemniscate. The energy flows with no beginning or end. We can once again use the analogy of the musical conductor. Think about the figure eight pattern the conductor so often traces with his hands and baton in the air as he directs his orchestra. From his vantage point, the conductor is able to control and move the energy of the orchestra using this symbol. When viewed vertically, the same symbol emerges as the caduceus, the staff of mystic wisdom carried by Mercury. Both mainstream and alternative medical facilities display this emblem. The caduceus symbol presented as intertwining snakes on a staff signifies the life force that travels through the human spinal column. The snakes represent a positive male current and a negative female current. When the two unite, the power begins to flow, providing the body with the circulation of life energy.

Additionally, the number Eight has been called the number of justice because it can be divided evenly into two Fours. The Fours can then be divided evenly into two Twos. Finally the Twos can be evenly divided into two Ones. The Four, we recall, represents regulation and order of the material world. When the Four is doubled, it repeats this command to gain control of elevated states of consciousness. With the influence of Eight, it is possible for individuals to realize their potential by learning to travel between the material world and the divine realms. The crossing point in the figure Eight is the eye of the needle that all persons must pass through when transiting between the two worlds.

Churches often incorporate the symbolism of numbers into their construction. When surveying the baptismal fonts in cathedrals in both Europe and the United States, we find that the fonts are traditionally octagon shaped. When the sacrament of baptism is administered, the octagon shape acts as a reminder to the inner self that our human spirit has the ability to regenerate itself in heaven or on earth. Baptism reaffirms our status as eternal beings who will one day make a shift to higher realms. The center of the infinity symbol (Eight) is the portal through which the transition to other dimensions takes place.

ILLUSTRATIONS

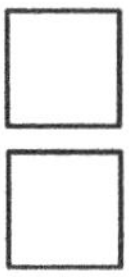

1.

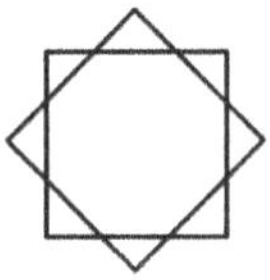

2.

3.

8

4.

1. Eight is divided equally into 2 squares
2. The squares interact to create an 8 pointed star.
3. The lemniscate or symbol of infinity
4. Modern 8

DESCRIPTIVE ADJECTIVES

Courageous, Forceful, Prosperous, Influential, Ambitious, Persevering, Powerful, Strong, Organized, Assertive, Commanding, Compelling, Successful, Dynamic, Industrious, Responsible, Businesslike, Capable, Principled, Disciplined, Self-Reliant, Energetic

EIGHT SPEAKS

- *When I open the morning newspaper, I usually scan the front page, then turn to the business section.*

- *It frustrates me to watch others struggle through tasks that I know I could perform easily and efficiently.*

- *I bought this house as an investment; however, I also expect it to serve my family's needs for many years to come.*

JOB PLACEMENT

Financial Planner, Commercial Developer, CEO, Professional Athlete, Professional Coach, Business Owner, Corporate Planner, Senator, Representative, Publisher, Bishop, Commissioner, Manager, Judge, Executive, International Law, Engineer, Stock Broker

CONTEMPLATE

Eight in the Bible

The eighth chapter of Genesis reveals that the floodwaters receded so that Noah and his family could finally leave the ark. We learn that God saved a total of eight human souls from the deluge. There were Noah and his wife, and Noah's sons Shem, Ham, and Japheth and their wives. These eight people were responsible for regenerating humanity. After the rains had ceased, Noah waited seven days, then sent out a dove. When the dove did not return, he knew that it was safe to leave the ark. The return to land commenced on the eighth day.

Note: Eight is the number of regeneration. Noah and his family served as human instruments of expression of the divine Eight energy.

A GLANCE AT EIGHT

Mythology figures:	Zeus, king of the gods, Nike, goddess of victory, and Themis, titan of justice & order
Zodiac Sign:	Capricorn
Tarot:	Strength
Gender:	Masculine-Feminine Balance
Mind:	Unified Conscious and Subconscious
Temperament:	Mental
Color:	Gold
Energy	Powerhouse

The Humanity of Eight

The Controller

I have chosen three people that I know personally who were able to deliver clear insights into the workings of the Eight vibration. They all maintain high profile positions, so I have identified them solely by their job titles. The perceptions these people shared with me were drawn from their unique life experiences. Although they come from very different walks of life, they each expressed a common message concerning the basic nature of the Eight vibration. None of these influential people seemed to be overly concerned with money or power, and they all brought up the subject of balance. As you read their personal commentaries, you will notice that each individual made it a point to emphasize how important it is to keep things in equilibrium. I am reiterating this essential detail because numerologists typically speak of Eight as the master of the material world and neglect this vital perspective. Eight is often type cast as a moneymaker and the one who can afford to buy the best that life has to offer. While this is generally accurate, it tells only half of the story. When we encounter people who are highly influenced by the Eight vibration, it is true that they usually appear to be prosperous and in control of their world. However, the complexity and spiritual depth of these people frequently remains unseen.

The following narratives allow you to be privy to the thinking of people who are heavily influenced by the Eight vibration. As you read their individual perspectives, you should gain a more complete understanding and appreciation of the principles of the number Eight.

The Psychiatric R.N., Administrator

When someone presents an issue, I immediately connect with all of the pieces to that person's puzzle. I see the good and the bad of his situation and all of the probable choices the person can make along with the ramifications of those choices. It is very hard for me to patiently stay in the moment with the client. I find it frustrating to watch him struggle while knowing I could easily solve the problem he is presenting to me; however, if I want to be fair to him, that means I have to keep my balance. He may have a lot of questioning, probing and analyzing to go through in order to ascertain a solution to his problem. It is not up to me to take away his process of discovery. I am only there to manage. I'm the administrator.

In my personal life, my take charge attitude frequently comes forward to solve problems. I think I get involved with projects because I recognize that they need done, and I naturally organize the people, places, and things to get affairs moving. I don't always feel that I need to be the person in charge, but I often find myself in that position. My call to leadership results from seeing what needs my attention and then stepping in to take care of things.

The Priest

I am very goal oriented. I find that if a person wants to achieve something, it is vital for them to keep focused on the goal. As a priest, I was taught that the purpose of our life here on earth is to get to heaven, and I maintain that goal for all of my parishioners.

I often find that people lack success at whatever they are trying to accomplish, and they end up feeling frustrated. The problem is that they get hung up on all the particulars and loose sight of what it is they are trying to achieve. When they are not dedicated to a plan, people get sidetracked easily and loose their sense of balance. I, too, can get distracted with the particulars, but when I notice my inattention, I bring myself back into focus on the goal.

In my profession, the ultimate end is heaven, and there is a formula that spins out of that objective. We are provided with a means and that is to follow the Ten Commandments. People need to stick to the formula.

For the past 30 years I have served as a Pastor and that position certainly has a lot of responsibility. There is a concern about money in this occupation because the bills have to be paid to keep the physical plant operational and all the different programs running. I have managed to retain a healthy viability in the parishes where I have worked. The logistics have to be handled. I take care of that so I guess you could call me the CEO, if you want to. It is an essential part of the job, but my focus is, and always has been, to get everyone of my parishioners to heaven.

The Telecommunications Specialist

Ambition is a funny concept. It frequently gets equated with chasing after money, but ambition can be used for many reasons other than obtaining more money. I like to think ambition takes you in the direction you place your passions. My drive is very subdued and sometimes I think something is wrong with me because I don't have ambition to go after money. I know that I have the capability to make more of it (money), but I don't. Don't get me wrong. I certainly understand that I need money to support myself, and I want to be comfortable with my life. I do want to have enough, but I am not concerned with becoming a business magnate. I have what I need. You see, it's not about money but about keeping yourself in balance.

I suppose I could become a millionaire like so many of my friends. There is always the choice to be made between time and money. A person cannot have both. I have a friend who was an artist by profession, and he drove a school bus to make a living.

His family circumstances were such that he knew that one day he would likely inherit a lot of money but he did not overly concern himself with that thought. Well, the day finally came that he did receive his inheritance. I ran into him after that time. He was no longer struggling with his art; in fact, he was jetting off to his show opening in Paris. He mentioned some investments to me that he was considering.

My friend used to have time, but now he has money.

Question: Are you saying that somebody like Bill Gates, who is presently considered one of the richest men in the world, does not have time?

Bill Gates doesn't have time. He may have houses and servants, etc. but he cannot stop thinking about money. In that way, he doesn't have time. I want more time.

Question: What will you do with your time when you get more of it?

Answer: *Nothing. I'll do nothing. I may just sit on the porch.*

The Philosophy of Nine

COMPASSION

It has been a long journey since our departure from the unity of number One. Through the study of numbers, we observed the development of the inner nature of mankind. We saw the creation of the family and the hard work it takes to build and maintain security. After basic needs were ensured, mankind then progressed further through learning and the expansion of the heart. On the seventh step of this process, when the work of creation ceased, we finally had time to rest, integrate all of the experiences, and appreciate the progress that we had made. After this period of repose, the number Eight regenerated the evolutionary movement. Eight renewed our focus and set things back into motion by coordinating the energies of all of the other numbers. Functioning as the administrator, Eight assured us that all tasks would run smoothly and that we would achieve our goals.

By opening to the philosophic meanings of numbers, we have been guided to a pinnacle that affords us an opportunity to expand beyond our familiar parameters. Our development has prepared each one of us to affirm with love:

I ACCEPT MYSELF AND ALL THAT I AM.

As we open to this self acceptance, we need to firmly acknowledge the existence of an invisible, unknown side to our being. The force of the number Nine makes us acutely aware of both our conscious and subconscious experiences. This expanded awareness allows us to gradually overcome the sense of duality, defined as the lower and higher selves. Nine purposefully moves us towards this great awakening so we can embrace the many facets of our being. As we study this final single digit number, we see that Nine readily incorporates the essential vibrations that have brought us to this apex. Nine loves all, accepts all, and serves all. It engages the necessary energetic qualities in the complete set of numbers to further our evolutionary progress and does so with love. It generously forgives mistakes that have been made along the way. Nine draws our attention to the awesome responsibility we have been given to care for and manage the extraordinary creation we inhabit. Through this grand and humbling task, Nine shows us that the environment reflects the condition of our soul. Nine teaches that when we nurture and advance this world, we are tending to ourselves.

It takes abundant heavenly light to elevate the whole of humanity above and beyond the limits of ordinary thinking. Nine provides us with that enlightenment. This vibration allows us to gain a multidimensional perspective of ourselves so we can fully appreciate who we are and how far we have come on the evolutionary path. Under the guidance of Nine, our inner vision combines with love. This advancement attunes us to our potential and the infinite possibilities available to us. We readily feel the compassion Nine emanates when we are under its influence. Generously imparting light and kindness without reservation, the mature Nine redefines success in terms of love and selfless service to humanity.

Nine can be illustrated with three equilateral triangles. Each of the triangles symbolizes a vital part of the three-fold nature of humanity: body, mind, and spirit. When the power of Nine takes control, these individual facets of our being can no longer function in a disjointed fashion. Body, mind, and spirit must align, creating a perfect reflection of one another.

Nine is the force that unites all the other numbers. Mathematically, if you add 1+2+3+4+5+6+7+8 it equals 36 or 9 (3+6=9). 360° denotes the circumference of the sphere. This may be the perfect time to bring to mind the expression *we have come full circle.* The circle (or cosmic egg), a symbol for God, signifies the place where we began, and Nine completes the journey by returning us to that original point of debarkation. Our numerical study has progressively outlined man's activities during his descent into matter. Nine completes the circuit of our study as it returns us to God. Nine functions as a unique looking glass by reflecting the invisible, our immortal self, in human form. If we count by nines, from nine to ninety, we clearly see a numerical pattern of reflection in the series set before us. (ie: the first number, 09 is reflected with the last number, 90, 18 is reflected with 81, 27 is reflected with 72 etc.)

09-18-27-36-45-54-63-72-81-90

Nine highlights the spirit or invisible part of our being and reminds us that the physical body passes away, but the invisible part of our being continues intact.

Fittingly, in mathematics, Nine acts as an invisible number. If you add Nine to any other single number the number will return to you unaffected. Examples follow:

9+5=14 (1+4=5), 9+7=16 (1+6=7), 9+2=11 (1+1=2)

Besides functioning as an invisible number in addition and subtraction, we find that it returns to itself in multiplication.

9x3=27 (2+7=9)

Although Nine's position ends a cycle, we find its immortal seed hidden in the beginning of the next cycle. Like the horizon, it serves as the ending point and the starting point of the beyond. A gestation period of nine months must be endured for a human baby to be born, and then life continues in multiples of nine-year cycles after birth. We can progressively map out the stages of life using a system of nines. Let's begin with the symbol for the number Nine, the three triangles.

For the purpose of following the cycles after birth, we will assign a period of Nine years to each of the individual triangles in our illustration below. The three main segments of life are built by adding another set of triangles to each progressive division.

(3x9=27) Age 0-27 is the first segment of life – Youth, Growth Years. ▲▲▲

(6x9=54) Age 27-54 is the second segment of life – Adult, Productive Years. ▲▲▲ ▲▲▲\

(9x9=81) Age 54-81 is the third segment of life – Retirement, Enlightenment Years. ▲▲▲ ▲▲▲ ▲▲▲

Man's progress forever revolves around the cyclic nature of Nine. We recognize its significance for humanity because it governs our evolution. Known as the number of service, Nine embodies the pure idealism responsible for originally projecting man's spirit. Emotional highs and lows are accentuated as the Nine undergoes a gamut of experiences from the mundane to the spiritual. While progressing through these events, Nine learns to forgive itself and send love to all parts of its being. Because Nine practices love, understanding, and forgiveness of itself, it is able to extend its sympathy, compassion and understanding to others. This charitable vibration constantly searches for altruistic works to perform.

Some say, the number Nine is as elusive as the horizon. We can eternally move towards the horizon, yet never come to own that space. Try as we may, we can no more capture the total meaning of Nine than we can reach the horizon.

ILLUSTRATIONS

1. 

2. 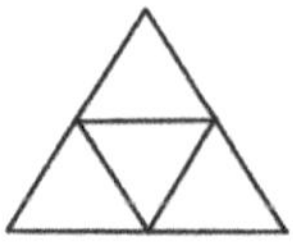

3. 

4. 9

1. three equilateral triangles
2. triangle within a triangle: Three equilateral triangles are set together to form a triangle within a triangle. This presents the eternal, watchful eye of God.
3. man offering his heart to God
4. modern nine

DESCRIPTIVE ADJECTIVES

Forgiving, Broadminded, Idealistic, Selfless, Sympathetic, Generous, Understanding, Compassionate, Loving, Humane, Hospitable, Impressionable, Demonstrative, Intense, Charitable, Emotional, Sensitive, Passionate, Intuitive, Enlightened

NINE SPEAKS

- *Each one of us embodies a small fragment of the infinite love of the universe.*
- *How may I be of service to you?*
- *I find narrow-mindedness impossible to fathom.*

JOB PLACEMENT

Rehabilitation, Philanthropy, Artist, Rabbi, Priest, Minister, Nun, Spiritualist, Healer, Mortician, Social Worker, Psychologist, Peace Keeper, Guidance Counselor, Writer, Director of Continuing Education, Herbalist, Teacher, Hospice Care, Recreation

A GLANCE AT NINE

Mythology Figures:	Apollo, god of healing, light, and truth, and his twin sister, Artemis, goddess of childbirth
Zodiac Sign:	Sagittarius
Tarot:	the Hermit
Gender:	Masculine-Feminine Balance
Mind:	Superconscious
Temperament:	Intuitive
Color:	White
Energy:	Magnetic

The Humanity of Nine

The Visionary

It's a daunting task to capture the expansive nature of a Nine personality. I asked Carol, a friend who has more Nine influence than anyone I have ever encountered, to help me with this personality. Carol and I met during our sophomore year in college, and have managed to keep in touch over the years. She currently holds an administrative position at a local university. One evening we met for dinner, and I asked her to describe how she perceives the overwhelming Nine influence on her life. This is what she had to say:

I have always felt alienated from people and the various cultures I encounter. I feel disconnected. It's like I am slightly outside of everything that is going on. It is very hard. All of my life I have felt like an outsider. I have one insatiable interest, and that is to explore the nature of our existence. I want to know God and why we are here. This is where my thought constantly goes.

Thank God for my job at the college because it forces me to keep myself grounded. As an administrator, I have to keep myself and my staff on track. The job keeps my attention on the very real concerns of the students. It requires me to focus on the here and now. If I don't have that focus, my mind just drifts. It is very difficult for me to stay with the concerns of the mundane world. It never ceases to amaze me how other people get so engrossed with things. They get all wrapped up in cars, shopping, hobbies and television shows. I guess these things represent their passions, but ultimately they have no meaning. Why do they waste their time? In the end, what is the real value of that item you took to the antique show? This is the way my mind thinks, and as a result, I am always looking to the spiritual for answers. I have never been able to enter fully into life the way that other people do. This is probably why I have felt like an outsider. Maybe I should have been a nun. I guess, in that profession, I would at least be with like minded individuals.

Don't get me wrong, I am a social person. I do enjoy being with other people and talking with them. I was at a picnic just yesterday and I had a good time. There are days however, that I need to be alone. I am planning to spend the holiday tomorrow relaxing and doing nothing. I don't always have to be at a party.

In my numerology practice, I have found Nines to be an enigma when it comes to socializing. They seem to easily attract a lot of people to themselves. They are caring and always make an effort to extend a helping hand. However, once they make a connection, they often pull back. The Nine becomes easily overwhelmed with the endless needs of other people, setting off a pattern of warming up to folks, then withdrawing. When Nine starts to feel the slightest energy drain, it scrambles for its refuge: privacy. In these instances, it may appear to the other people that the Nine extends an invitation, and then later rescinds.

Notorious for their sensitive nature, I asked Carol to speak further about Nine's emotional nature. Her reply follows:

I am always feeling the tug of my personal feelings and it often becomes a burden. I think this is how I often get sidetracked. I cannot just cut through an issue and complete a task without taking the emotional trip. My thoughts get clouded easily. It's a pain. However, because I experience the emotional expression qualities so readily, I think I have become more sensitive to the needs of other people.

A NOTE ABOUT THE HUMANITY OF NINE

Recently, I had a woman in my office who was trying to explain her approach to her work and relationships. She talked a lot about her feelings and the difficulties she has experienced. I listened patiently while staring at the many Nines in her numerology chart. When she had nearly exhausted her anecdotes, I interrupted and handed her the manuscript narrating Carol's thoughts concerning the Nine experience. Carol gave such a well thought out, accurate description of the persona of Nine. I could see that this client was quite impressed with Carol's insights. After taking it all in, she said she was speechless. It amazed her to read such an accurate version of her own perceptions of life.

Sometimes people with Nine energy complain that they get tired of being in the role of service to others. Knowing this, I strive to be kind and accommodating to those people heavily influenced with the number Nine. In catering to Nine types, I have learned something about their basic nature. Although they may complain about constantly being the giver, they feel uncomfortable in the role of the receiver. They specifically search for and seek out the downtrodden so they can be of service to those people.

If you want to befriend a person with a Nine persona, you must remember to keep your distance. They don't like to feel smothered. Always be sure to give them enough space so they can maintain their privacy. Nines are best paired with other Nines. In these relationships, they can relax and feel assured that a Nine like themselves, will not likely be placing demands on them.

ILLUSTRATIONS AND KEY WORDS FOR NUMBERS ONE THROUGH NINE

MALE

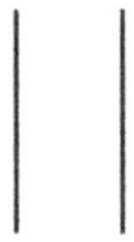

FEMALE

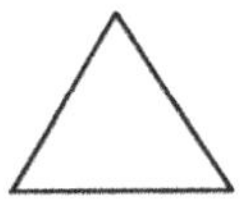

CREATIVITY

REALITY

ADVENTURE

BALANCE

THOUGHT

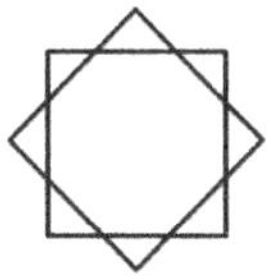

MOTION & CONTROL

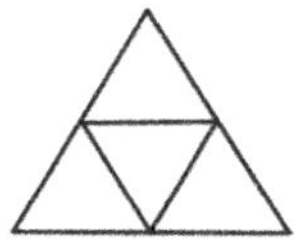

EYE OF GOD

Men and WoMen

A NOTE ABOUT THE MASCULINE AND FEMININE ASPECTS OF THE NUMBERS 1 THROUGH 9

All of the numerology books that I have come across agree that the number One symbolizes the masculine principle and the Two corresponds to the feminine. Many systems continue the progression by having the sexes take turns in their assignments ie: Three is masculine, Four is feminine etc. Some systems refer to this exchange as a natural flow of positive and negative energies. They use the two rotating charges to explain that like ocean tides, at times your surge of energy becomes heightened and at other times it wanes. After study and experimentation, I have found that the rotation system does not make sense; therefore I do not accept this formula.

The Philosophy of Number section in this book encourages an appreciation of the roles the sexes play in regard to numerology. I think it is important to review the gender aspect of the numbers one last time as understanding the influence and placement of the masculine and feminine principles is essential to numerology.

In the last section of each philosophy next to the gender

sub-heading, you will find that the numbers One through Five (with the exception of Two) are all masculine numbers. The numbers Six through Nine show their genders to be balanced. When referring to a number as balanced, I mean that the feminine principle has reintegrated itself into the unified force from where it originated. Two is the only number that is purely feminine. As you will recall, the Two made the sacrifice of separation from the One solely for the purpose of objectively studying the self. The female stands alone in the symbol of Two. She does not particularly relish her position but willingly takes on the duties of an observer while waiting for an opportune time to reunite with the One. The feminine principle is not looking for its turn to take charge of the creation. Females do not seek to usurp the power of the male.

If men on this planet would only come to realize that they are not in competition with women, perhaps they would stop treating women as if they are a lesser species. I, along with the vast majority of American women, am only marginally suffering from this problem, but I know that there are many women on this planet who are grossly suffering. The shocking yet glorious revelation that we must all face to stop the age old battle of the sexes declares that **there is only one sex.** THERE ARE ONLY MALES. FeMales are Males that carry the additional power of insight. **We are all men**. (WO)MEN ARE MEN. Yes, we do have different reproductive organs so that we can further the species, but we are all men-- **WE ARE ONE.**

Returning to the Philosophy of Two, remember that it was man's idea to study himself. He sent part of himself out to get a better view of things, and that part of himself is called the woman.

Now, when that part returns to him with the data it gathered, he seems to have forgotten that long ago he ordered this job to be done. He begins treating the returning part of himself as the enemy and seeks to overpower it. Men need to remember who the women are and listen to them(selves). Women were sent out to complete a very important and specific mission. Through years of practice they have refined their skills and they have become very good at their task. Women constantly gather information and knowledge. They infuse their findings with insight to deliver wisdom to our culture.

When we accept the premise that **WE ARE ONE,** we can then understand how each of us possesses a masculine and a feminine nature. As we continue to grow, each one of us must get in touch with these two aspects of ourselves. We must accept that Women have the capacity to take up leadership positions, and Men have the capacity to serve as househusbands, and vice versa. When we deny this range of flexibility, we diminish our vast potential and our culture suffers as a result.

Generally, it seems a majority of men do not like to look at themselves or consider advice from others. I make this statement because I have counseled countless women in my numerology practice who came to discuss marital problems. These *wo*men almost always arrived at my office alone. The effectiveness of the numerology counseling is reduced when only half of the marriage is present, so inevitably, I would ask the woman if her husband might be willing to see me or go to a traditional marriage counselor or psychologist. I rarely received a positive response to this question. Perhaps these men don't understand that learning to face themselves and accept the insights that their women can give them is the key to making them stronger and more powerful. Although a man has the peace and happiness of a good marriage to gain, his independent nature is usually threatened when faced with accepting women.

If men could only see past this threat, then they would finally realize that ***she is he.*** When our group consciousness acknowledges the fact that **WE ARE ONE,** then we will be able to enjoy a new paradigm. Until that time the battle of the sexes pointlessly continues.

Part II

APPLICATIONS

Putting the Philosophy to Work

You now have in your possession many of the powerful truths numerology presents. This language, which also represents a philosophy, conveys information about some primal concepts that have occupied human minds for ages. It encourages a contemplative thought process.

Age old questions concerning our existence emerge within the study of numerology. **Who am I? What is my purpose?** Answers to these profound questions can be easily deciphered using number codes. When people begin to take notice of the meanings of numbers, they quickly see that the answers are packaged inside their own name.

The very personal application of the language of numbers is quite captivating, and it motivates many people to pay attention to this subject. When I give my introductory lecture to a group, I always watch the faces of the audience and observe their expressions becoming progressively more attentive as the relevance of numerology begins to dawn on them. Once people realize they can apply the meanings of numbers to their lives to understand their own behavior, they become fascinated with the subject.

Early in my career, I was asked to make a presentation to a group of professors and students at Carnegie Mellon

University, a very prestigious institution in Pittsburgh. This opportunity indicated to me that I was following the correct path. Nevertheless, I was quite nervous and feeling a little insecure as I launched into my impassioned speech about using the number set as a language. The lecture hall was very quiet and my audience seemed to be listening intently, however, half way through my talk, a professor, situated in the center of the room, stood up and began to climb over the legs of everyone seated in the row in order to make his way to the exit. He was obviously in a hurry to depart. My first thought was that he must disagree with my perspective on numbers. I felt especially bad about him leaving so abruptly because of where he was seated and became concerned that the rest of his colleagues might follow suit. I stopped my lecture and in a friendly way I said to him, "It's too bad that you have to leave because I'm just getting to the best part."

He looked at me squarely and said, "Oh, I'm not leaving. I have to make a call to have someone cover my next class so I can stay and hear more." This reply, spoken by a mathematics instructor at Carnegie Mellon University was quite the compliment to me, that young numerologist so many years ago. As we approach the applications now, I realize that I am probably at that same point that I was in my lecture when the professor excused himself. It is time to take our fascinating study of numerology and apply it to you.

Our first step in personalizing this knowledge will be to examine the numerical foundation of our alphabet. When we open to the relationship of numbers and letters, we see that the correlation makes it quite easy to translate any word into a series of numbers. If the target of our investigation is the self, which is usually the case, then we must consider the name. Transcribing the name into a sequence of numbers is the key to getting started. In learning to interpret these numbers (and letters) we will be analyzing the cornerstone of the given name. The cornerstone is comprised of the first consonant and vowel. We will then count how many ones, twos, threes, etc. are included in the first, middle, and last names. This tally forms a report called the Karmic Record. Finally, we will use the information gathered in the Karmic Record to study the physical, emotional, mental, and spiritual make up of ourselves or the person we are charting.

Traditionally, when composing a chart, a numerologist will ask for the original name given at birth and the subject's birth date. It is important to use the official given name when making calculations even if the client has never used that name. During my many years of practicing numerology I have found that observing this rule is absolutely vital. The following two examples, taken from my private practice, beautifully illustrate the validity of this tenet. These case studies exemplify the powerful impact that the given name creates.

Lana, a client in her fifties, has worked with me for the past seven years on many different issues. When Lana first came to my office, she confided in me that she had been adopted when she was two days old. Although her new parents kept the first and middle names she had been given, they changed her last name to their own. The surname of her birth parent was never used by Lana for any reason. During her formative years, Lana was aware of the fact that she was adopted but gave little thought to the details of her history or her original given name.

In her twenties, Lana got married and began using her husband's last name. So when she made her first appointment to see me, she was actually on her third last name. I noted all of the pertinent information Lana gave me during the interview. But when it came time to do the numerology calculations, I just used her original surname. During our reading I gave her very precise, in depth descriptions of her personality, her innermost desires, and her approach to problem solving. I wanted her to carefully discern whether my verbal illustrations of her were correct. At the time, Lana wasn't familiar with numerology, and remained unaware that I based my calculations on her original surname. My perfectly accurate descriptive reading of her personality, motivation, and temperament left Lana amazed. Since I have had the opportunity to work closely with Lana for many years since our first meeting, I have been able to personally observe the precision of her numerology based on her given name at birth.

My second noteworthy case concerns a woman in her thirties called Kate. When Kate was born, her mother named her Agnes. Shortly after her birth, her mother filled out the necessary paperwork and the hospital staff mailed the documents to the state department. Two hours later her father arrived at the hospital to find out that he had a beautiful little girl named Agnes. This was a bittersweet moment. He was proud and happy with his new daughter, but also angry that his wife had named the baby Agnes without consulting him. Every time he heard the name Agnes, it grated on his nerves. He flatly refused to accept his wife's choice of a first name for their daughter. So, on the same day, within a few hours of registering the original birth certificate, the parents mailed an amended document. When Kate told me this story of being named twice within a three hour time period, I really wasn't sure what data to use to calculate her numerology chart. I asked her if she would work with me on this project and we would study both names together.

What we learned was that each of these names expressed extremely different qualities. We found that the name Agnes (along with her middle and last names) expressed qualities gifted to an actress. Agnes gave her a talent to dramatically communicate the emotional side of life. Agnes had the potential to become a star. Kate, on the other hand, exhibited a conservative, shy, and studious personality. Lacking the flamboyant style of "Agnes," Kate had to be sure of her facts and figures before speaking up.. After many hours of discussion concerning her numerology, Kate admitted she could feel both of these characters inside of her while she was growing up. She struggled continuously between the two energies, and then finally, in her late teens, she felt the Agnes energy waning. By her early twenties she began to feel comfortable with the sole vibration of Kate, and Agnes seemed to be gone for good. Kate found her niche as a translator and an English teacher in Europe. I accept the dominance of the Kate energy, however, at times, I still recognize the occasional outbursts of the actress vibration. We closed the case, and have not been in touch for awhile, so I don't know if Kate wants to acknowledge her hidden talent—Agnes.

Occasionally, I have a person who comes to me to ask advice about changing her name. She will narrate her reasons, and proudly reveal the new name she has chosen. After hearing her explanation as to how she feels the change is likely to transform her life, I ask if she has any way of erasing the old name, as this is something people don't usually consider. Taking a new name is acceptable. I'm not against engaging a new vibration to uplift the spirit, but what does happen to the original name when we do this?

In my experience, the original name does not disappear, and this presents a problem.

I often liken the process of taking on a new name to throwing stones into a pool of water. We toss the first stone then watch the ripples progress out from the center of the splash. When we throw another stone into the pool, it produces another set of ripples, but the second set of ripples does not negate the motion of the first set. The second splash may in fact create conflict with the waves that were first put into motion. At the very least, we can expect this action to compromise the power and complicate the motion initiated with the first stone.

Another analogy I like to use to help people understand the consequences of a name change is the decision to dye one's hair. Coloring the hair could potentially have a very big impact on a person's life. That person may be treated differently by others after making this change, Although altering hair color from brunette to blonde may conjur diverse reactions, eventually the natural color of the roots will emerge. Hair color changes that are made can only be superficial.

We can correlate the hair dyeing experience very nicely to the implications of modifying a name. A person may create a savvy new name, and the new name could potentially have a powerful impact on her career or the type of friends she attracts into her life. She may feel fulfilled and happy with the new character she has created, but she still has to contend with the part of her that she has masked. The vibrations of the original given name will continue to make their way to the surface in the same way that the natural color of the hair grows back. The birth name projects the character and personality of each person and covering up those traits does not mean they have disappeared. Friends who pay close attention will notice those qualities, and the individual will continue to connect with those deep aspects of self.

There is no simple way to erase the effects of the original name

Sometimes people reject their given names because they do not understand and appreciate the unique role they have been assigned. On an unconscious level, some individuals are simply overwhelmed with the work they perceive it will take to live up to the full potential of their given names. This in turn can make them angry at their own vibrations. Imagine a young child, dressing up with his father's hat and tie. When that child steps into a pair of Daddy's shoes, his little feet look lost and he doesn't have the necessary control to walk properly. The spirit of an individual can experience a similar event when it moves into a new name. A good and perfect name can often challenge the spirit. A name can be experienced with an overpowering, confusing, or weighty sensation. In most cases, the name suits the person and all he needs is a little time to grow into his power. We can all think of a name that had little appeal to us, until a movie star or some other famous figure appeared with that moniker. Once the name glowed in the spotlight, we began to accept and maybe even like the name that at first seemed so distasteful to us.

Sound advice for anyone who may be thinking about changing his or her name would be to first seek to understand the beauty and the power of the original given name. People commonly change their names because they cannot live up to their name's inherent power. They willingly settle for a name that has less command. Think about how we tend to soften the names of young children. Robert becomes Robby or Bobby. Deborah becomes Debby. It often takes time for children to grow into their names, and it may also take time for adults to accept the power imparted by the young one's name.

A classic example of the progressive acceptance of one's name, for both the individual and his contemporaries, resounds in the story of Fred Smith. Throughout his entire life, Fred Smith was employed by a local manufacturing firm. Beginning in high school (ages 16-18) he spent his summers working at the plant. At the time, people knew him as **Freddie,** the mail boy. While Freddie attended college he continued his part time employment. During those years, he became a clerk in the shipping department and was referred to as **Freddie Smith** (ages 19-22). He was well liked by the people he worked with, and he appreciated the employment opportunity. After graduating from college, the company offered this hard working young man a full time job. He eagerly accepted the offer and at age 23, Fred embarked on his career as a supervisor on the loading docks. He was given a decent starting salary, his own office, and a plate on the door that read: **Fred Smith.**

After a few years working on the docks, a position in the controller's office became available. Fred bid on the job and began his upward move in the company. His new business card announced: **Fred T. Smith,** purchasing agent. This was a very nice (and well deserved) promotion for Fred. At age 26, not only did his career begin to take off but he was dating the boss's daughter as well.

Five years pass before we check up on Fred once more. During that time he married his sweetheart and began working closely with his father-in-law in the executive suite. The placard on his desk read **Frederick T. Smith.** The years continued to roll by, and in due course his father-in-law announced his decision to retire. His father-in-law's retirement meant that Fred would assume the prestigious position of Chairman. When this transfer of power was about to take place, the company newsletter printed a touching article about the accomplishments of this hard-working father and son-in-law team. The article affectionately referred to the new Chairman of the Board as their very own **Frederick Theodore Smith.**

The Progression to Frederick Theodore Smith

Freddie	Mail Boy
Freddie Smith	Clerk
Fred Smith	Foreman
Fred T. Smith	Purchasing Agent
Frederick T. Smith	Executive
Frederick Theodore Smith	Chairman of the Board

As chairman of the board, his distinguished and commanding name, Frederick Theodore Smith, flows off the tongue with ease. When he served as a mail boy however, that same name would have been difficult for him and his co-workers to accept. It can take some time to grow into a name that carries as much power and authority as Frederick Theodore Smith. We occasionally encounter children who have dominant and extensive formal names, and we do feel comfortable using their given names. Be watchful of those little ones, as they will one day rise to greatness.

Reviewing Numbers 1 through 9

1 - I AM

2 - I AGREE

3 - CREATE

4 - I CONFORM

5 -I EXPERIENCE

6 - I LOVE

7 - I THINK

8 - I CONTROL

9 - I FORGIVE

The Relationship of Numbers and Letters

It is impossible to learn to read without first knowing the alphabet. Think about how we present those initial language lessons to very young children. We don't just spill the alphabet blocks out on the table with the attitude: Here are the letters kids. Go to it! That approach would never do. A crucial element to the learning process is becoming familiar with the sequence of letters. Each letter has a specific placement beginning with A, the first letter, and continuing to Z, the twenty-sixth letter. Do you remember learning the alphabet song? It serves as an innocent way to teach children the proper succession of letters as they relate to the number set. When singing the alphabet song (or signing our name to a legal document), the proper sequence of letters along with their hidden numerical counterparts proves to be extremely important. The number set supports the alphabet, and individual letters express hidden meanings of equivalent numbers. All written and verbal communications reference our numeric structure and meanings encrypted therein.

Table of Numbers & Letters

1	**2**	**3**	**4**	**5**	**6**	**7**	**8**	**9**
A	B	C	D	E	F	G	H	I
J	K	L	M	N	O	P	Q	R
S	T	U	V	W	X	Y	Z	

Sequentially, J appears as the tenth letter of the alphabet and falls under the number One in the above table. If we reduce 10 by adding the numbers together (1+0=1), we arrive back at the number One. S represents the nineteenth letter of the alphabet. After we reduce 19 (1+9=10), the sum 10 can be reduced again (1+0=1) returning us to One. As you explore various applications of numerology, you will need to refer back to this table of numbers and letters.

The Cornerstone

The first simple, yet profound, application we will make is to correlate the meaning of the numbers to the first consonant and vowel of the given name. When we refer to the vowels, we are speaking about the vibrations contained in the sounds of the letters A, E, I, O, and U. Y assumes the vowel status only if there is no other vowel present within a syllable. Together, the first consonant and vowel hold tremendous potential energy. We have the ability to animate that potential with our voices. As we initiate their sounds with our vocal chords, we activate the power contained in these symbols. Creating these initial sounds aloud can be compared to striking a match and igniting a fire. The resonance is transformational. The engagement of the leading sounds of the name immediately enlivens that person or business.

Using the letters of the cornerstone, you can predict (or anticipate) how a person will react to the many events and influences in his or her environment. The impact of these letters cannot be overstated. The very first letter that appears in the original name, whether it is a consonant or a vowel, presents the dominant vibration. If the name that you are studying is Frances, the Letter F dominates and the qualities of the A are directly incorporated into this power sound. If you are considering the name Evelyn, the power of the vowel will take precedence and the qualities of V will work to enhance that energy.

There are many names that begin with blended consonants. Examples would be Charles, Thelma, Charlotte, Phillip, and Shirley. The cornerstone here gets a bit bulkier when the sounds of two consonants merge. When studying such names, I advise you to take all three letters into account (the first two consonants and the first vowel). The H, although hidden or blended in these names does pack a punch. A woman named Thelma, for example, may at first appear to be shy and demure (T=2 influence) but as we get to know her, we find that Thelma has a businesswoman concealed under the surface of her sweet facade (H=8 influence). She can put on quite a display of force when a situation demands it. Never underestimate the concealed power of a Thelma or anyone else with a hidden H in their cornerstone.

Note: There a quite a few names that have two consonants before the vowel comes into play; examples would be Crystal, Clarence, Steve, and Florence. In these names the consonants are not blended, therefore the cornerstones remain C-y, C-a, S-e, and F-o respectively.

There are also names that carry blended vowel sounds (diphthongs) in the cornerstone. Two examples that come to mind are the names George and Louis. As a rule, when you have two vowels placed together, the vowel energy becomes greater than the consonant, even though the consonant precedes the vowels. To put it simply, the vowels outnumber the first consonant. The cornerstone of the name George conveys energies pulling in opposite directions because of the diphthong in this name. Here we have an E=5 that wants to explore and test the limits, placed with an O=6 that begs to nestle at home and be the responsible parent. As a result of this configuration, the name readily manifests anxiety, and often causes people with this identification to feel they are literally being pulled apart. Friends and associates of a person named George will frequently comment that he easily loses focus and scatters his energy. When I stand next to someone with this first name, I can perceive the waves of anxiety he emits. The name gives off a nervous and jagged vibration, infiltrating and breaking up smooth and seamless harmonics. Parents should be made aware of the effects the eo diphthong can have before naming their children. Anyone who has the responsibility of naming an infant (or even a pet) would benefit from learning the basics of numerology.

The name Louis presents us with three sequential vowels. Only two of the vowels, however, are part of the cornerstone, as the the letters I and S form the second syllable of the name. The O=6 is creative, artistic and home oriented. The U=3 is creative, artistic and energetic. We are able to readily see that the energies of O and U have some common ground with which to work. The L of Louis is an expressive and communicative vibration as is the U. Never give confidential information to a Louis and expect him to keep quiet! It is not a defect of character that he cannot keep a secret, rather it is part of his basic nature to express.

It would be wise for you to take time to examine your own first consonant and vowel. Use the Table of Numbers & Letters to identify your cornerstone numerically and then review the meanings of those numbers. Be sure to read the influences of the dominant letter first. (Remember, this can be either a consonant or a vowel, depending on which comes first in the sequence.) You can use the philosophic information outlined in part one of this book or look to the brief synopsis that follows.

Keep in mind that we are beginning with one influence. As you continue the study of numerology, you will learn how to use many more pieces of the puzzle. This information will help you get started in making accurate and objective assessments of yourself and others, including family, friends, and even rivals.

The Power of the Cornerstone

Names that begin with A, J, and S:

Boldness and creativity glow in the aura of these outstanding people, and they make a strong impression in any crowd. Like clever magicians, these imaginative individuals quickly materialize ideas, to make their wildest dreams come true. They skillfully get their way because they usually know what they want, and they're not afraid to go after it. Being both ambitious and determined, these authority figures refuse to take *no* for an answer, and rarely entertain compromise. An inherent talent for leadership dominates as they thoroughly believe in themselves.

Names that begin with B, K, and T:

These very gentle souls can be easily crushed and are frequently moved to tears. They intently seek the approval of other people and are willing to compromise their wants in order to satisfy others. Infinitely patient and always willing to wait their turn, they keenly observe and study the mistakes of those who go before them. Minute details are scrutinized and remembered for future opportunities. Be careful what you say in their presence, because they are ***always*** listening.

Names that begin with C, L, and U:

Naturally gregarious, their copious imaginations are always brimming with potential. These people effortlessly produce an overflowing well of creative ideas that *must* be expressed vocally. In a classroom setting, they are often the ones that the teacher must continually ask to *"be quiet, please!"* They enjoy life to the max, and it shows by their youthful appearance. They love to laugh.

Names that begin with D, M, and V:

These honest, trustworthy citizens have a strong moral code of conduct. After making a commitment, these souls will go to great lengths to honor their promise. Representing the best of our work force, these everyday heroes constantly put forth their best effort to take care of the duties assigned to them. They are comfortable following the well worn path carved by the proven experiences of past innovative thinkers.

Names that begin with E, N, and W:

It's difficult to keep up with these fast past characters because they move around like a streak of lightning. Pinning them down is a nearly impossible feat. Just when you think you have captured their attention, they are off and running in another direction. Constant motion and engagement in mental brainteasers keeps them comfortable. Being fast, witty, and interested in everything, these people become a source of information for others. This group is truly *in the flow.*

Names that begin with F, O, and X:

Responsible is the perfect descriptive adjective for this group. When everyone else runs off to play, you will find them in the house cooking, cleaning, organizing, or preparing for another day. A profound need to be surrounded with the comforts of home generally inspires them to create visually pleasing and relaxing environments. Motivated by a strong sense of duty to care of other people, they often get involved in planning and cooking meals for friends. These nurturing souls love to share themselves through counseling and their conscientious natures endow them with an ability to listen on a very deep level.

Names that begin with G, P, and Y:

These starting vibrations give us life long students --those inquisitive souls who never stop asking *why*. They are the observers and scholars who take the time to deeply examine issues that come before them. Science and philosophy prove to be two of their favorite areas of study because research is the focus of activity in both of these fields. As their questioning minds mature, these people usually become specialists in their chosen fields. Quite commonly, the pursuit of truth and knowledge will permeate leisure activities as well as their vocations.

Names that begin with H, Q, and Z:

The concept of responsibility becomes greatly expanded with this group. These individuals feel as if they were born to take charge of the masses. They are more than willing to take up the reigns of planning and organization, and are not afraid to crack the whip. While keeping their sights on the goal, they unceasingly prod the troops forward and guide them to victory. Being excellent money managers, it is not uncommon for them to play a major role in building a corporation and at the same time develop a personal fortune for themselves. These vibrations ignite true executive power.

Names that begin with I and R:

These are the idealists who believe in mankind. They always seem to show a generous nature, and manage to toss something into the donation basket, even when their own finances are tight. Being charitable with their time as well, we see them just about anywhere a volunteer is needed. They serve on political action committees, raise funds for school groups, push wheelchairs in hospitals, and usher in church. A generous spirit leads them to forgive, and to hope that tomorrow will be a better day.

Subtle Tones of the Alphabet

Referring back to the Table of Numbers and Letters, it is important to consider some subtle variations in the letters of the alphabet. Following the number system, there are 3 lines on the table. Line 1 includes the letters A through I. Line 2 covers J through R, and line 3 covers S through the letter Z.

THE LETTERS OF LINE 1
A THROUGH I

Imagine these letters as being brilliant and striking in appearance. If represented by color, they would deliver an explosion of primary hues similar to the bold reds, yellows, blues, and greens we remember from our kindergarten crayon boxes. All color emits frequencies of power, and the energy that appears in this first row of letters is robust. This first line represents the youngest letters of the alphabet. As such, they have yet to prove themselves to the world. In their exuberance to assert their authority, they usually have a tendency to be overpowering. These letters delight at being in command and feed on the attention they draw to themselves. They will take every opportunity to demonstrate how special they truly are.

THE LETTERS OF LINE 2
J THROUGH R

The second line of the alphabet emanates maturity and confidence. This group doesn't feel the need to constantly broadcast authority and draw attention to themselves. Occasionally the urge to make an impression surfaces, and when it does, you may catch them primping in front of the mirror in preparation for a grand entrance. Displays of arrogance are rare, however, and most of the time, you will find these letters to be accepting and comfortable with who they are.

Letters J through R thrive on respect from others. This need motivates them to work diligently at keeping their skills polished. Developing expertise in a given field not only provides them with internal satisfaction, it also earns them the admiration they subtly crave. On close inspection, you will find that their capacity does contain a restrained expression of superiority. If we were to identify this line of the alphabet with color, we would perceive a wash of bright and soft pastels.

THE LETTERS OF LINE 3
S THROUGH Z

These are the ancient letters. Experienced and wise, these characters don't feel the need to prove themselves to anyone. This last line of the alphabet can be compared to a class of people that we sometimes refer to as "old money". Most of the time, they appear to be completely self-satisfied and frequently keep to themselves. However, when the mood strikes them, they can be boisterous and overbearing and will eagerly adorn themselves in gaudy colors and accessories in an attempt to show off their status. Just as they begin to draw attention with their outrageous behavior, these smug letters then return to the seclusion of self satisfaction.

Five of the eight letters presented in this line (U, V, W, X, and Y) have a cup or a container feature to their appearance. These specific letters often choose to allow their capabilities to sit unused in that figurative urn. The colors projected by the letters S through Z are the muted, darker hues.

In Conclusion...

The numbers one through nine work in conjunction with the letters of the alphabet. Each number vibration undergoes a slight variation when it appears on lines 1, 2, or 3 of our alphabet table. (9 is the exception to this rule as it only appears on lines 1 and 2.) This placement allows for subtle variations in the expression of a specific energy. The letters A, J, and S for example, are all number One vibrations on The Table of Numbers & Letters. Each of these Ones, however, produces a slightly different style due to its placement on line 1, 2, or 3. We can readily distinguish a shift of the One energy when examining the impact of the first letter in the cornerstones of Amy, Jim, and Susan.

Amy

People whose names begin with A always tend to be a little self-centered. Those who carry the name Amy follow this rule, making a strong and lasting impression on others. Her fashions reflect her character with bold choices of color and design. With her unique and dramatic style, Amy poignantly announces to the world that she is number One. She wants things done her way and expresses her demands in a direct manner. There may be times that she will politely ask for an opinion, but she often disregards the information she receives. Amy only pays attention when another person's point of view benefits her exclusively. Her number One status keeps her focused on satisfying her own desires and compromise is not part of her vocabulary. Note: The name Amy has a long A pronunciation rather than the short A that begins the name Andrea. It is the long A that exaggerates all of the positive and negative attributes of number One.

Jim

A person using the name Jim tends to be little more easy-going than someone whose name begins with A. In the competitive arena, it appears that he is not as vocal and demanding as an Amy. Jim operates at a more casual pace and incorporates a sense of receptivity to the ideas of people around him. He shows a willingness to sincerely listen. Afterwards, he will sort out the things that are not important to him and offer to compromise on those issues. Although his self centered motives are toned down, don't let his approach fool you. This clever competitor also enjoys being number One, and he plays to win. The game is most enjoyable when Jim is in the lead.

He's not consumed with being first, but he bears that number One need to be *first-class.* Jim dresses to impress Jim. Although he is not overly concerned with drawing attention to himself, his impeccable style usually generates quite a bit of recognition and appreciation from others.

Susan

The name Susan also begins with a number One letter, but as you may recall, S is one of the ancient letters. People with names that begin with S to feel that they are Sages. Susan believes that she has all the answers. If you should disagree with her, she will suggest that the two of you sit down to "discuss" your differences. Prepare yourself for a one-sided negotiation, for she can converse for hours while blatantly ignoring your point of view. Susan is watchful and patient when it comes to getting her way. She will sit at any boardroom table for the sake of debate, but like a shell game, she will cleverly move the conversation around until she has others who sit at the table convinced that they want things her way.

Susan often goes for the retro styles because she has a yearning for the old way of doing things. Fashion trends do not excite her because she keeps her sights on classic designs. She is secure in knowing that the classics forever remain number One.

The Karmic Record

The Karmic Record is a cumulative report that registers all the learning that you have amassed during your many lifetimes. Since it is cumulative, it allows us to view the complex influences gathered in the unique life you led prior to your birth into this present lifetime. This collection of information reads like a history book. It highlights your strengths and directs your attention to those areas that need further work.

In a professional reading, the Karmic Record is one of the first areas I like to review. I will usually ask the client to imagine herself standing on the stage of life. I then tell her that I am going to place a backdrop behind this stage. I want the client to view the information of the Karmic Record as one would view painted scenery on a theater stage. It is an essential component of the stage, but not the focus of attention. It passively occupies the background yet creates perspective for current activity. The backdrop contains pertinent symbolic clues that the client can freely associate with her current real life experiences. It's important that she reflect on this imagery and decide if she resonates with the information. When she is able to acknowledge some of these past pieces of the puzzle, she can begin to better understand her own passions in the role she now plays.

This is how I would like you to understand and use the concept of the Karmic Record. Past lifetimes should not be the focus of our attention, but they do deserve our recognition. We live in the present, but we build upon the lessons inscribed in our past actions. The joys and sorrows we have accumulated here at the earth school are filed within the Karmic Record. The past (whether it be positive or negative) colors our present endeavors. This influence can be very subtle. We may not even realize that we are being impacted by our connections to a forgotten era. Difficult events that took place in a past lifetime can create an overbearing mood of sorrow in the current lifetime. True healing requires that, in some capacity, we recognize past mistakes. In some cases, that awareness can be accomplished abstractly through dreams. After enough information concerning an issue gathers in the subconscious mind, the current personality has an opportunity to release from the energy of that issue. The better we become at this process of forgiving ourselves and others, the more mindful we become concerning the choices we are presently making. Another gift that evolves, as we discharge ourselves from the past, is the development of compassion. The compassion we have for ourselves and others keeps us focused on the greater good, and motivates us to live more creative and productive lives.

Although an infant has not had time to accumulate life experiences, it carries a Karmic Record outlining its talents and weaknesses. Objective observation of a young person as she grows into adulthood validates the accuracy of this record time after time. We unequivocally know that babies come with innate abilities and deficiencies at birth. Where do their talents and imperfections originate? Logic brings us to the conclusion that the newborn earned (or failed to earn) these credits in a previous lifetime. This premise provides us with the basis for the Karmic Record.

To create a chart of your own Karmic Record simply write down your original given name on a piece of paper. Assign the correct numerical value from the Table of Numbers and Letters in the space below each letter, then, count how many 1s, 2s, 3s, 4s etc. you has accrued in your name. After you have determined how many times each number is represented in your name, add up the column. An Example follows:

Calculating a Karmic Report for

Mary Katheryn Woolworth

M a r y	K a t h e r y n	W o o l w o r t h
4 1 9 7	2 1 2 8 5 9 7 5	5 6 6 3 5 6 9 2 8

1. 2
2. 3
3. 1
4. 1
5. 4
6. 3
7. 2
8. 2
9. 3

21

The sum of the column in our example is 21. Double check the number of letters in the name with this sum to ensure that you have not omitted any letters. There are 21 letters in the name Mary Kathryn Woolworth and so the column should add up to 21. If the totals do not match, scan the name again to make sure you have not overlooked a letter and check the addition of the column. After you have verified the correct total, it is time to study the Karmic Record.

Understanding the Karmic Report

We will look at the numbers accrued in the name of Mary Katheryn Woolworth and highlight some of the features of her report. Remember, the karmic report serves as an evaluation of Mary's past. The sum total of this report shapes not only her physical form but it influences how she thinks, feels, and intuits information as she continues to practice her current life lessons. The report tabulates the many features Mary has learned in previous incarnations. Viewing her report, we readily observe that Mary has accumulated abundant knowledge. All students of the earth school periodically serve as teachers at some level, and it appears that Mary is ready to share some of the wisdom she has amassed.

As you are learning to work with the karmic report, you will practice using both left and right brain functions simultaneously. Since people are taught to almost exclusively use left brain logical thinking in school, they need to be reminded to re-engage the right brain, allowing it to scan and wonder. This special task requires you to be both logical and contemplative as you approach the statistics we have gathered concerning our subject, Mary. It's easy to do this by pretending to have a private conversation with her. The object of this exercise is to allow your intuitive thoughts to surface as you converse with her. After the process is complete, you can then analyze your deductions and make conclusions. When you work with your own experimental subjects, I would advise that you relay your findings to those people and ask if they resonate with your hypothesis.

As you follow my stream of logical and contemplative thought, you may want to refer back to the calculations in the Karmic Report for Mary Katheryn Woolworth. After reading my musings concerning Mary, I recommend that you begin studying your own karmic record. Remember to let your mind wander as you go through this experimental exercise. It will be interesting to see what develops as you have an imaginary conversation with yourself.

A Sample Stream of Logical and Contemplative Thought

Mary, I see that you have accumulated three credits in quite a few of the earth school subjects. You must have been very creative in your last lifetime. The triads are grouped under Two, Six, and Nine. I also see that you have the number Two listed in categories One, Seven, and Eight. You have a very organized karmic report. The average person carries only one Two in his name and you have three of them. That indicates to me that you have been a very creative woman in your past lifetime. I can see that you are obviously strong in the subject of marriage because you have three Sixes in your name. The average person has only one Six. Eight is abundant on your report also. You have two number Eights. When I consider that many people have zero Eights on the report, two is a lot. Two Eights tell me that you have established and controlled the material plane quite nicely in a previous lifetime. It can sometimes be surprising to a person to learn she has to earn a living again in this lifetime, when she has already proved herself to be adept in this matter.

In diverse civilizations throughout the world, wealth is commonly measured by the number of children a person has. Given that you are a woman now, and you have lived as a woman just prior to this incarnation, and you have an abundance of sixes (the domestic vibration), I think it is safe to entertain the notion that you probably had many children in that past life. Materially speaking, it is likely that you were quite comfortable in that lifetime. A person who has the configurations that you have here, could have been a rancher with a lot of land surrounding her, or she could have been a city dweller with a lot of money in her pocket. I will have to let my intuitive self take over to get a better feel as to what the nature of your wealth was all about, Mary. My inclination is that you were the rancher type. I think you worked very hard in that previous life. Anyone who creates a lot of children is likely to have a lot of work caring for those children. You have only one Four in your name and that is average. You do however have four number Fives in your name. The Five on the report can be mental or physical. Because you have four fives, I believe you were involved with a lot of hands on type labor. I think you had babies, washed clothes, raised chickens, and canned vegetables. You put a lot of power into that home and marriage. If you don't want to get married in this lifetime I could understand why. Then again, you probably loved every minute of that former existence and would repeat it again in a heartbeat.

All of the higher number vibrations are very well represented in your report. You well surpass the average in subjects Six, Seven, and Eight. You are on the high end of the scale with number Nine. I also want to note that you are not missing any numbers. This indicates to me that you have lived here at the earth school in many past lifetimes and gathered a lot of experience in the advanced subjects. A full chart like yours makes others appear to be only beginners! I often wonder why people return to this earthly existence when they seem to have mastered all of the subjects on their cosmic report card. I think you are here Mary, because some members of your tribe or family still have work to do. You are here to tutor, and help them along with their lessons. If you are living in a family situation now, where the family members are as well developed as you are, then we should consider that this family reincarnated in order to work together on a large scale project. Advanced souls sometimes reincarnate in groups to effect changes politically, artistically, socially, or spiritually. Being advanced doesn't mean that one gets to lounge around, for there is always work to be done. When we are experienced, the occupations just become more sophisticated.

I think it is important that I review the lesson of Seven with you Mary. You have two Sevens on your report. The Seven you recall indicates deep thinking and meditation. When I see two or more Sevens on a report, it tells me that the person was most probably deeply religious in a past lifetime. Having even two Sevens can indicate that a person lived a lifetime in religious service. I would like to ask you, Mary, if you consider yourself to be particularly religious in this lifetime? My guess is that you are not likely to be interested in religious services. In fact, you may have an aversion to attending these Saturday- Sunday gatherings. If I am right about this, the distaste you have for religious services might have you wondering if there is something wrong with you. Why don't you enjoy going to church like the rest of your community? I am here to tell you that you have graduated from that subject! Once a person has graduated from any institution, (ie: high school) she can appreciate what she learned there but she is not likely to want to return and repeat the classes. Mary, if you do feel inclined to further yourself with more Seven lessons, I would recommend in depth studies on the spiritual path. Your interest may take you to something that is not at all religious in nature. You may want to investigate yoga, biogeometry, some form of meditation, astrology, or numerology. There are a host of spiritual pursuits to explore that have nothing to do with religion.

In summary I would like to say the following about your most recent past incarnation:

You were not poor.

You produced a fantastic family.

You lived a very full life.

I get the feeling that you were an exceptional person.

You must have had a lot of influence over other people.

Points to Consider

THE AVERAGE NAME CONSISTS OF

APPROXIMATELY 19 LETTERS.

LETTERS	NUMBER/MEANING	FREQUENCY
A, J, and S	1--ACTIVE	appears 3.6 times
B, K, and T	2 - PASSIVE	appears 1.3 times
C, L, and U	3 - CREATIVITY	appears 2.9 times
D, M, and V	4 - PRACTICALITY	appears 1.1 times
E,N, and W	5 - EXPERIENCE	appears 4.0 times
F, O, and X	6 - LOVE	appears 1.0 times
G, P, and Y	7 - ANALYTICAL	appears 0.95 times
H, Q, and Z	8 - AUTHORITY	appears 0.90 times
I and R	9 - COMPASSION	appears 2.5 times

- *One* represents the ego and gives a person a sense of identification. It is highly unusual for this number to be absent from a chart. When many Ones are present on the Record, it can mean that subject of your study has been quite busy fulfilling his own basic needs. The number One embodies strong masculine energy. If a woman has an excessive amount of Ones in her chart, consider that she may have lived as a man in a previous incarnation. Likewise, consider the possibility that a man with more than the average number of *Twos* in his chart may have lived as a female in a past incarnation.

- The creativity expressed by the number *Three* is found on almost all karmic reports. Three experiences are nearly impossible to avoid. When we say "hello" to our neighbor we use this spontaneous energy. This number displays a healthy presence because all people naturally express their creativity as they go about the business of living their lives.

- If *Four* is missing on the report, don't assume that a person was unproductive in past lifetimes. He could have contributed something extraordinary without being physically involved in the task. After a person has spent a lifetime doing mental work, the opportunity to apply himself with basic manual labor may be a welcome relief. In the evolutionary scheme of things, practical, mechanical jobs are no less important than mental work.
- *Five* is represented more than any other number. Since this figure endures as a symbol of man, it corresponds to the thinking mind, and the sensing body. It is extremely rare to have Five missing from a name. Five represents learning that comes from direct experience, and most people take this route before developing more sophisticated means of gathering information. An excessive number of Fives displayed in a name indicates that the person has probably developed

strong sensual appetites. Many Fives can also be an sign that the person survived previous incarnations by making due with what they could eek out of life on a daily basis.

- Notice that the higher numbers (6 through 9) appear less often than the lower numbers (1 through 4).

- The principle of balance that naturally manifests in the number *Six* takes an extraordinary amount of practice to achieve and maintain. Most people work on attaining this point of satisfaction through their relationships. Marriage, parenting, counseling, and healing are perfect outlets for the Six energy. There are many souls who have completed this assignment in previous lifetimes but continue to engage in this labor of love. Even when people have many sixes in their names, they will often enjoy running through this lesson again.

Seven and Eight on average, appear less than once in any given name. As you study various karmic reports, you are bound to find people who are missing these numbers. The absence of any number does not indicate that the person has failed a subject. Perhaps she has not yet taken that specific course during her progressive incarnations. If only one number is absent on a person's report, that missing principle will stand out like a sore thumb, and the person often feels compelled to pursue the lesson the missing number represents

Sample Karmic Report Cards & Short Commentaries

Please Note: **Calculations were based on the complete names; however, only first names appear in the sample.**

Mona

1	4
2	0
3	0
4	2
5	5
6	2
7	0
8	1
9	0

COMMENTARY:
Mona has a lot of work to do in this lifetime. She has been heavily involved with basic survival and family issues in previous incarnations. Now it's important for her to focus on her creative insights. She needs to enjoy life in order to reach her full potential.

John

1	2
2	1
3	1
4	1
5	3
6	5
7	2
8	2
9	1

COMMENTARY:
John's energy is like a waterfall that cannot be stopped. He loves and cares for others. he has been a physician for many lifetimes and is likely to continue on this path of caring.

Flo

1	0
2	1
3	5
4	0
5	7
6	2
7	0
8	1
9	2

COMMENTARY:
Flo is extremely sensitive and caring when it comes to others. She has given of herself with reckless abandon in past lifetimes. Now it is important that she learn to include herself in this formula.

Robert

1	2
2	1
3	9
4	4
5	4
6	0
7	1
8	1
9	3

COMMENTARY:

Robert is extrememly creative. He was an accomplished artist in past lifetimes. In his current life, he is free to pursue any subject that captures his imagination. Robert is exceedingly capable.

Jesse

1	2
2	2
3	5
4	0
5	6
6	2
7	2
8	1
9	1

COMMENTARY:

Jesse is a brilliant individual. He must work at shifting his numerous creative ideas from his imagination into material reality. he is slow and overly careful in his approach.

Andrea

1	7
2	1
3	1
4	2
5	7
6	0
7	0
8	0
9	1

COMMENTARY:

Andrea is self-centered. she lives inside her own head. Balance is the most important of her many lessons in this lifetime. She can accomplish the balance lesson quickly with the right marriage partner, but it won't be easy for her.

Elements of Expression

The Elements of Expression are tied directly to the information that we gathered from the Karmic Record. By studying the Karmic Record we were attempting to learn something about the energetic influences carried from past lifetimes. Observe humanity and you will find that we use a distinct, repetitious, evolutionary pattern. We faithfully structure the current reality upon the past. In terms of numerology, the figures found on the Karmic Record are destined to shape the Elements of Expression. Scan the Karmic Record thoroughly before you begin a Numerology reading for anyone. I don't recommend that you dwell on the topic of past lifetimes, but I find it vital to understand something about the person's history, for he builds his current life on that foundation.

Think of the Elements of Expression as the tools we presently use to create our reality. The prototype of our earthly dimension embraces a four fold pattern, which is unique to the inhabitants of this material plane. Four distinct aspects of existence manifest in every human being. These modules consist of (1) The Physical Body, (2) The Mental Body, (3) The Emotional Body, and (4) The Spiritual Body (our inner knowing). These four components structure our existence and are described in numerology as the Elements of Expression. By projecting the self through these channels, we can focus our awareness and manifest our desires. Through these expressive outlets we communicate, love, and create our lives.

Specific numbers encapsulate the energies of the physical, mental, emotional and spiritual planes of our existence. I outline these groups of numbers along with a descriptive analysis of each category in the following pages.

Mental Numbers

One, Five, and Eight

These numbers drive us to construct, explore, and manage through mental applications. Number One offers the constructive mode of thinking through the initiation of new ideas. Five uses its communication skills to take the mind exploring. The Eight oversees and manages the mental operations that the One and Five set into motion. This trio facilitates the release of divine creativity and its movement into form. They energize every thought with the properties of willpower and resolve to ensure that our inspirations and theories have a proper chance to manifest in the world. The mental channel exemplified by invention and problem solving is One, Five and Eight's area of expertise. These numbers constantly work to balance our intuitive nature with reason and logic.

Emotional Numbers
Three, Six, and Nine

These three numbers overwhelming rely on the energy of the heart. The Three freely expresses emotion. Six balances that expression, and the Nine releases it at a higher octave. The emotional energy discharged by these numbers is like a mist moving both inside and outside the boundaries of the physical body. Feelings and imagination flow and expand along the subtle communication network created and maintained by Three, Six, and Nine. These numbers provide us with a valuable sensitivity that links all of us together with invisible threads. Our innate ability to share emotions allows us to access the feelings of other individuals. Emotions free us from the lengthy analytical processes of the mind and provide us with heart to heart communion. Our ability to have empathy for one another reinforces the concept that *we are all one.*

Physical Numbers
Four and Five

The number Five literally ignites the five senses, our channels of sight, sound, taste, touch and smell. Five compels us to learn through direct experience, and it immerses us in the physical world in very sensual and sexual terms. This force makes us true explorers. Communication and direct interaction are the hallmark of Five energy.

The number Four also takes the hands on approach, but it is much more practical. It is not driven by curiosity, like the Five.

A sense of duty motivates the Four. It feels an obligation to take care of essentials, and it accomplishes this by efficiently organizing and regulating all the particulars in its reality. Four radiates a very down to earth, matter of fact, attitude in its management style.

The energies of these two numbers counterbalance one another as they team up to provide our physical needs. Five works to expand our horizons while Four manages and preserves our reality.

Spiritual Numbers

Two and Seven

Two and Seven are considered numbers of inner knowing because they have the ability to turn our focus inward. These energies look fearlessly into the depths of the soul in search of answers. The inward motion of the thought processes is sometimes referred to as right brain thinking. (Left brain controls reason and logic and right brain serves the intuitive functions.) When thinking takes the form of prayer, meditation, or contemplation, it is able to flow inward and access information directly from Spirit. These modalities form our spiritual channels, and many people regularly access their higher power through these avenues.

Since Two and Seven work to facilitate our soul connection, people whose names contain a lot of these numbers have a tendency to develop strong extrasensory powers. Extrasensory, of course, refers to the senses that reside beyond the physical body. These senses operate at a different frequency and serve a spiritual purpose. Both Two and Seven encourage us to recognize our relationship with the divine and seek enlightenment. To aid in the accomplishment of this goal, these energies consistently take time to express appreciation for All That Is.

The Dual Nature of the Number Five

In the above review of the mental, physical, emotional, and spiritual properties of numbers, Five is categorized as being both mental and physical. It earns a dual listing because Five contributes to the operation of both of these avenues of expression. Most numerologists seem to overlook this fact and erroneously designate Five as a representative of physical expression only.

In the language of numbers, Five is known as the number of man. Imagine, if you will, a stick figure of the human body showing a head and a trunk with arms and legs extending outward. This simple yet powerful symbol of man displays both his mental (head) and physical natures (trunk with arms and legs).

Leonardo DaVinci gave us his famous drawing of a man standing in the center of a circle with arms and legs outstretched. The human depicted by DaVinci projects absolute perfection. In a perfect model, we would find man balanced in regard to mental and physical skills. During my many years of researching numerology, I have found that this equilibrium does not often occur. When there are many Fives present throughout the chart, the person will tend to either be very physical and sensual or will express a strong mental talent. I surmise that as he continues to evolve, man will learn to incorporate and balance both the physical and the mental aspects of his being into his human form.

Finding and Calculating the Elements of Expression

Having already learned how to determine the Karmic Record, this next phase of calculations becomes a simple process. Look at the record that you have assembled for yourself and add up the numbers that are assigned to the physical, mental, emotional and spiritual groupings. Remember to reduce the final sum to a single digit. Two examples follow. (Please note: Calculations were based on the complete names, however only first names appear in the samples.)

HELEN'S KARMIC RECORD AND CALCULATIONS

1	2	MENTAL POINTS - 1
2	1	Add the quantities assigned to 1, 5, & 8.
3	1	2+6+2=10 (Reduce the number 10)
4	3	1+0 = 1
5	6	PHYSICAL POINTS - 9
6	2	Add the quantities assigned to 4 & 5.
7	1	3+6 = 9
8	2	EMOTIONAL POINTS - 4
9	1	Add the quantities assigned to 3, 6, & 9.
		1+2+1 = 4
		SPIRITUAL POINTS = 2
		Add the quantities assigned to 2 & 7.
		1+1 = 2

STEPHEN'S KARMIC RECORD AND CALCULATIONS

1	6	MENTAL POINTS - 4
2	2	Add the quantities assigned to 1, 5, & 8.
6+5+2=13		
3	1	(Reduce the number 13.)
4	1	1+3 = 4
5	5	PHYSICAL POINTS - 6
6	0	Add the quantities assigned to 4 & 5.
7	1	1+5 = 6
8	2	EMOTIONAL POINTS - 4
9	3	Add the quantities assigned to 3, 6 & 9.
		1+0+3 = 4
		SPIRITUAL POINTS - 3
		Add the quantities assigned to 2 & 7.
		2+1 = 3

Interpreting Points of the Elements of Expression

You have yet another tool to use in your expanding work with numerology. After reviewing the interpretations that I have outlined here, take some time to examine how this information applies to the people you know best.

Practice on yourself, then expand your study to include friends and family.

Starting with your given name, calculate the points you have accumulated on the mental, physical, emotional, and spiritual planes. Don't force yourself to fit into every feature of my analysis. The explanations in this section are meant to provide you with a framework from which to start. You will find subtle differences with every person you examine.

As you are learning, it would be best to read and consider the interpretations as they apply to your friends, then interview those people. Discuss with them how these Elements of Expression manifest in their lives. They will be able to verify much of the information you gathered about them and enhance the data by giving you personal examples that corroborate the theory. As you go through this process, you will begin to understand your friends better while developing an understanding of these essential building blocks.

You may want to compare the number totals of your own physical, mental, emotional and spiritual expressions to the results you find in the chart of another. Note that when you share the same number of mental points with someone else, the two of you will tend to think alike. If you happen to have several of your aspects match exactly, the two of you will share a lot in common.

Mental Expression

1 MENTAL POINT

People with one mental point are extremely creative in their thinking and generate very unique approaches when it comes to problem solving. When working in a group, those with one mental point enliven others with many original, interesting, ingenious ideas. They show little interest in how things were done in the past, as they are always precisely focused on the ideas they wish to manifest here and now. They also have an uncanny ability to direct their attention and achieve their intended results. It's amazing though, how quickly their concentration evaporates once the inventive portion of the job is complete. As soon as these creative thinkers get the ball rolling, they eagerly pass the maintenance chores to others so that they can continue engaging in exciting, inventive tasks over and over again. Working as part of a think tank would be ideal for this type of mind.

The natural leadership capacity exhibited by a person with One mental point manifests because of his strength of will. His determination drives society forward as he plows into the unknown with reckless abandon. The One thinker considers himself to be the center of the universe. With this premise in place, he fully expects others to serve his needs. His ultimate plan is to take care of himself first, and over time, allow the benefits of his self centered strategy to trickle down to his followers.

2 MENTAL POINTS

The person with Two mental points enjoys a slower pace than most people . Even when it comes to making simple decisions, he needs time to process his thoughts. You will find him willing to listen intently to what you have to say, but don't expect a quick response. Typically, he formulates an answer a day or two after he has had a chance to mull things over.

Having Two mental points engages the right brain and enhances intuitive thinking. When you pin this person down for a logical explanation, his reaction may be to stare blankly at the clouds. Just when you are ready to give up on getting a reply, he suddenly blurts out a response. The intuitive faculties are so integrated that sometimes even he is not aware of how he derives his brilliant conclusions. If you happen to work with someone having Two mental points, I recommend that you periodically question him thoroughly. The person with Two mental points is apt to retain useful information, however, they are not inclined to divulge what they know unless they are asked (or invited) to do so.

The Two, you will recall, is very detail-oriented. Its powerful and subtle influence makes a person inclined to absorb abundant conversational facts and environmental details previously during his encounters. For example, a man with Two mental points may sit in a meeting and never take a single note, yet he will be able to repeat nearly every word that the lecturer had to say as well as comment on his attire. Along with his amazing ability to remember everything that was said, he will also be able to tell you what the previous presenter wore when he spoke at the same event last year!

3 MENTAL POINTS

People with Three mental points are easily distracted. This mental expression can be in the middle of unfolding an important undertaking when something grabs its attention and it decides to go skipping off in another direction. Scattering one's energy is a hazard that results from being both highly creative and multi-talented. Three mental points takes the mind is here, there, and everywhere. It creatively works like a little spider spinning all kinds of designer webs. On a good day, it can churn out more creativity than it can ever possibly hope to manage.

Although this thinker enjoys reading and writing, the Three rejects the tedium of taking traditional approaches to problem solving. It would much prefer to invent its own solutions, rather than engage in boring research. When in doubt, the Three mind finds resolution by discussing the challenges it faces. It loves to talk, and this natural tendency leads the Three to think aloud and engage others with questions.

At one time or another, we have all heard an encouraging instructor say that there is no such thing a foolish question, however, this mind defies the professor. People with Three mental points love attention and enjoy making a game of formulating confounding questions. They have the wit of a proud comedian and the Three allows them to weave humor into all kinds of serious moments. Most of the time these friends are delightful, but occasionally their immaturity takes over and it can be unbearable. Although spontaneity is a wonderful gift, it can be stressful to watch them tackle a project when they have not made the proper preparations. Unfortunately, there are times that others refuse to take them seriously because of the innocent childlike thinking they project.

4 MENTAL POINTS

Every professor loves the student who strives to meet his standards, and the person with Four mental points makes it her goal to do just that. Her obedient mind seems to have inherited a talent for following the rules imposed by the academic community. She excels at following instructions and approaches all subjects with order and discipline. Beginning on page one, she reads every text book cover to cover. In the classroom, her methodical note taking shines. This student firmly believes in making a detailed outline that highlights important points of interest. Math often becomes a favorite subject because she thoroughly enjoys the routine of logically following a sequence of steps to arrive at an answer. Classes in this field of study provide her the opportunity to work within her comfort zone.

The student with Four mental points learns effectively when ideas are compartmentalized and labeled. A "hands on" approach to learning often proves best with her because she is not an abstract thinker.

Total confidence in her logical and vigilant mind leads the person with Four mental points to believe that her method of thinking is the way a person is *supposed* to think. Although she can be quite content adhering to the traditional way of doing things, she does have a desire to contribute a piece of her own original thinking to the body of knowledge created by earlier scholars. As she cautiously takes this step, it is important that she be able to reference her ideas to the standards already in place. She is not one to adventure alone into totally unknown territory, and becomes uncomfortable if asked to explore too far outside the box.

5 MENTAL POINTS

Five mental points describes what I refer to as the electric mind. People who possess this type of intellect can read and process new information as fast as a streak of lightening. Their fascination with speed often develops into a technology fetish. This makes them love to experiment with electronic equipment so they can quicken the flow of new information. They enjoy acquiring every new electronic communication device that can be found on the market. Keeping current with cutting edge technology often becomes their primary focus.

People with Five mental points are very adept when it comes to multi-tasking. Since these thinkers spend a great deal of time on the phone and online, they usually find ways to combine their communication opportunities with manual tasks that also need to be done. Interestingly, their comfort level increases when doing several things at the same time, and they get bored when they find themselves attending to only one task. Remote control is commonly regarded as the special tool that enhances their already quick abilities.

It may come as a surprise, but these technology wizards can also be avid book collectors. They find it comforting to surround themselves with knowledge and work hard to fill their bookcases with subjects of interest. I often hear them admit that they don't have time to read all the volumes they have collected, but just knowing that the books are available brings satisfaction. People with Five mental points have a voracious appetite for knowledge that can sometimes get out of control. When this happens, their energy becomes scattered, thinking becomes shallow, and they lose focus on current needs. They must be vigilant about exhausting themselves in this way.

6 MENTAL POINTS

People with Six mental points have a strong connection to the emotional field. This interrelationship of mind and emotions manifests as great sensitivity in their thinking. In fact, separating feeling and thinking is nearly impossible for them. The emotional component usually takes them deeply into an issue before they become motivated to engage the mind. This precedence makes it seem that their minds cannot function without their emotions. The strong feelings that commandeer their thinking usually create an obstinate nature. Once people with Six mental points form opinions, all the logic in the world will not sway them.

People with Six mental points are always subconsciously focused on the safety and protection of their loved ones. They enjoy planning events where they can draw friends and family together and entertaining often becomes their way of caring for others. Hosting parties provides them with opportunities to dispense their well intended council. You will find them at these gatherings circulating and mentally probing their guests in a kind and thoughtful manner.

The imaginations of people with Six mental points revolve around creating the right atmosphere for every holiday or special occasion. Romantic ideas often fill their heads, diverting their attention to fantasy shopping excursions when they should be dealing with reality. They can spend countless hours thinking about how they can redecorate their homes to make them more attractive and comfortable for the next time they entertain guests. Domestic projects including gardening provide a great outlet for their artistic thinking and loving natures.

7 MENTAL POINTS

The mental expression of number Seven describes the quiet philosopher. Here is the person who is willing to ponder a myriad of questions and answers in a search for deeper meaning. He finds it impossible to accept superficial explanations and continues to ask "why?" until he gets in touch with the fundamental principals for which he searches. This is the mind of the determined student, the scientific researcher, the philosopher, the priest, or the professor. Anyone who carries Seven mental points requires calm and quiet in his surroundings to carry out his work. The library represents a truly sacred space for him, and he thrives within its boundaries. He welcomes all opportunities for solitude which give him the time and space to listen to himself think. Since he doesn't feel a great need to be expressive and he doesn't typically invite others into his field, he can remain quite satisfied living a private and peaceful existence in communion with God for extended periods of time.

As it is commonplace for people with Seven mental

points to become lost in their own thoughts, their mental travels often replace the need for physical adventure. Using various forms of prayerful meditation, they can infuse their beings with tremendous satisfaction. Transcending the finest level of thinking to become one with the Creator provides them their greatest joy.

The sculptor Rodin captured the essence of this expression in his masterpiece The Thinker.

8 MENTAL POINTS

Eight mental points presents the administrative mind. This person can feel the power and promise of varied thoughts and immediately prepares to harness their potential energy. This keen investor doesn't waste time deliberating over small matters. Instead, he directs his energy toward studying the bigger picture. He realizes that he has been innately programmed to direct and guide the masses, and he intends to fulfill that mission. The person with Eight mental points has a destiny to take charge of other minds. His goal is not to usurp the power of the individual, but to focus on the best interests of the whole, be it his family, his company, his political party, or wherever else he places his allegiance. To better understand the influence of Eight mental points, consider the simple example given below that clearly illustrates how this energy manifests in a practical way.

A stay at home mother who has this type of mind would be concerned more about managing the needs of her immediate and extended family than she would be about completing daily household chores. She would rather hire someone to do those menial tasks because she is occupied with masterminding the next family reunion, contributing to school board committees, and completing other civic duties. She always keeps an eye on investments, and it would not be unusual for her to spend part of her hectic day trading stocks online. She is the bright lady who organizes car pools and baby sitting clubs in her school district. To better manage her time during the holidays, she coordinates relatives to each bring a favorite dish to the family gathering. This woman realizes the power of her own thoughts, and recognizes that she can increase that power by commanding the thoughts and actions of others.

9 MENTAL POINTS

It is a little challenging to describe how the person with Nine mental points processes information because her analytical approach remains very unstructured. A great deal of her thinking never leaves the subconscious realms, so at any given moment she may only be vaguely aware of what is percolating in her mind.

The person with Nine mental points constantly relates the personal events unfolding in her life to the broader perspective of the world in which she lives. For example, when making the decision to have a child, this woman takes into account how her pregnancy will effect the over all population of the planet. Caring for the environment is one of her major concerns and any issue that impacts the ecosystem automatically moves to the top of her list. Her expanded awareness and her love of nature motivate her to practice the belief that:

I am one with the universe, so whatever I do for the world, I do for myself.

This thought sums up the selfless philosophy of all thinkers who are influenced by the number Nine. They genuinely care about planet earth and tirelessly strive to expand on ways that we can better our lives here.

People with Nine mental points make excellent counselors and are naturally drawn to help other people with their problems. They have a true ability to understand other peoples' predicaments because they constantly maintain an expanded point of view. Nine thinkers readily offer empathy and comfort to all those who are open to receive. Tolerance and unlimited compassion are the special gifts this mind imparts to humanity.

Physical Expression

1 PHYSICAL POINT

He makes a strong statement with his brilliant fashion sense, and his inner confidence is candidly reflected in his bold appearance. The primary colors are appealing to this individual because the bright reds, yellows and blues clearly project the assurance he feels. Although it may not be his goal to be a trend setter, he frequently finds himself in that position. His unique choices intensely mirror the exceptional person that he is. It is supremely important to him that all of his goods and purchases be new and perfect. Perceived imperfection produces annoyance and frustration, and the slightest flaw can throw this person into a state of imbalance. His preferences seem to magnetically draw him to department stores that specialize in tailored clothing, and an attraction to expensive suits cannot be denied. The desire to put together a very sharp custom made wardrobe directs his shopping and compels him to stubbornly refuse to accept anything less than the best.

In his living space, he favors simple furnishings and styles that offer clean lines, along with pure and bright colors. He has an obvious need to reflect his unique tastes and original design concepts in his home. If possible, he should get directly involved in creating his personal environment. This person gains tremendous satisfaction when he can "do it himself," without any help from anyone else. It energizes him to tackle the many decisions that go into customizing his surroundings.

The strong masculine and ego-centric nature of the number One is unmistakably visible in all of the daring material choices that this person makes.

2 PHYSICAL POINTS

Appearance means everything to this lady. She carefully browses the fashion magazines and store displays to gauge what she is going to wear. Her intent, however, is not to dazzle anyone with the latest fad; rather, she only wants acceptance. Being received and loved by others is always her underlying goal.

Dark and quiet tones have an appeal to the person with Two physical points as these shades help to keep her world soft, quiet, and subtle. She doesn't reject bright colors entirely but prefers to use them sparingly. Lighter shades will often be used to accent her wardrobe. Highly critical of her own appearance, she dreads being placed in the spotlight when any detail is out of place. Getting caught wearing out of date eyeglasses for example would make her feel quite embarrassed.

The focus she places on physical impressions can cause anxiety even when she thinks about having others into her home. This unfounded stress makes it extremely difficult for her to relax in her surroundings. Until she finally realizes that she is her own biggest critic, she may never be able to truly enjoy entertaining others. Commonly, there exists only one very real problem with living space of a person with Two mental points and that is clutter. Parting with the many items she constantly accumulates proves to be nearly impossible for her.

The person with Two physical points must work to protect herself from the assault of the natural elements in this unruly world. Since she does not have a robust constitution, putting on the sunscreen, having only one glass of wine instead of two, and getting to bed at a decent hour are all very important to her. Two physical points usually expresses a delicate nature and health concerns must be closely monitored.

3 PHYSICAL POINTS

The people who maintain Three physical points display an overwhelming need to run and jump and play. No matter what their chronological age may be, these people always look and act younger than their peers. Their special blessing is an energetic nature and a youthful appearance.

They love to make a statement with fashion. You might find them shopping the "junior" department well into their fifties, as they find delight in modeling the ever changing fads. When they make an appearance, they usually project an upbeat and quite theatrical image. Wearing t-shirts with funny sayings entertains them, and they enjoy communicating with others in this playful way. These impish characters get a kick out of sending messages with fashion and accessories and take advantage of every occasion to express their feelings with different outfits. Halloween tends to be one of their favorite holidays because it creates the perfect opportunity to communicate with their playful imaginations.

People with Three physical points splash a lot of color into their wardrobes as well as their homes. Sometimes bright isn't bright enough for them. At those times, you will find them out shopping for neon. Having a voracious appetite for the latest styles, it would be prudent for them not to invest a lot in any particular theme for their living space. Flexibility is important for those who like to keep current with the trends. Obviously, people with Three physical points just love to shop. This tendency should not be interpreted as extravagance on their part. Their inherent creative urge constantly drives them to anticipate and search for inventive and fashionable designs.

4 PHYSICAL POINTS

This person is endowed with a strong physical body and blessed with good health. If he should come down with an illness, his body fights hard to regain its balance. He takes the sure and practical avenues of approach when he needs help. If his doctor advises a treatment plan, he will take his medicine exactly as prescribed. Sometimes, I compare his attitude about his body to the mechanical expectation he has for his car. When he turns the key, he presumes it should go without hesitation.

Clothing his sturdy and dependable frame presents an exercise in practicality. His tendency is to intentionally choose items that will serve him well for an extended period of time. It's important for him to take time to inspect his choices thoroughly to ensure the quality of the goods before making his final decision. As he stands at the register waiting to pay, he internally reviews each item's functionality and durability. He never seems to have a problem opening his wallet to buy functional necessities such as a new shirt or shoes, but he is rarely moved to extravagance. Luxury items usually don't resonate with his sensible and prudent approach to shopping.

His practicality clearly manifests when he shops for new

furniture. Traditional Furnishings is the department where you will find him. When looking for a new living room set, for example, it's important that he find a comfortable, functional couch. He won't fuss about superficial concerns such as the name brand or the colors in the upholstery. Essentially, he wants to be able to feel secure and relaxed when sitting down. Since his tastes usually stay within conventional parameters, the rooms of his house will generally be boring. Although it may seem that he is not expressing himself with his home furnishings, the standard motifs do convey his traditional nature.

5 PHYSICAL POINTS

Five physical points often produces hyper physical activity in a person. The individual expressing Five physical points needs to run, jump, play, and squish his toes in the mud. He craves sensuality and sexuality. Sweating reminds him how much he loves this physical existence. As the senses rule in his world, it can be challenging for him to control his appetites and often he feels overpowered with physical urgings. When this happens, it seems that his only option is to go with the whims of his senses. Overindulgence of any substance always looms as a temptation. Overeating or excessively using cigarettes, alcohol or drugs can easily overcome this person and cause real problems in his life.

Cautionary notes aside, his very physical nature has him leaping to his feet when there is a call for one more volley ball player at the company picnic. Because of his high level of activity, sportswear is always a good option for him. A wide variety of styles and colors converge in his wardrobe. It seems that he is able to get away with combinations that would never suit anybody else. The person with five physical points ranks as a fashion leader who frequently ushers in new styles. He is always on the move, anticipating the next trends soon to hit the market. When he decorates his home, his senses play a dominant role in his planning. As a result, his living room is apt to be filled with a variety of textures, colors, and the latest electronic equipment. Incorporating audio visual stimulation, to please himself and his guests, usually plays an integral part in the overall design.

6 PHYSICAL POINTS

Grace, beauty, and artistry permeate the body of the person with six physical points. Although we may have never met her before, we can expect her to be strikingly beautiful. She radiates a soft and gentle quality and has a loving touch. Typically, the qualities emerging in her physical body will also be found in her surroundings because our model strives to replicate her attractiveness in her living and working spaces. Her welcoming aura echoes in the vase of fresh cut flowers and the warm and inviting collection of furniture she has arranged for her guests. This wonderful hostess keeps a close watch on her house wares and pantry so that she is always suitably prepared to accommodate a special dinner party. Romantic lighting, candles on the table, and the glow of burning logs in the fireplace are never forgotten as she places the finishing touches on her environmental design.

A person with Six physical points presents her artistic side when she dresses, designs an interior, or plants a flower garden. Subtle shades and fine fabrics draw her attention as she carefully chooses balanced combinations for her wardrobe. Her mood, whether it be bold or sentimental will in some way be reflected in her dress. Both men and women who have Six as their physical expression have a strong connection to the energy of the heart. You can easily see the sensitivity of these people mirrored in their physical bodies. It's important to realize that emotional disturbances can intensely wear on their physical hearts. People with Six physical points must recognize this fact and protect themselves accordingly.

7 PHYSICAL POINTS

The person with Seven physical points gets anxious easily and even the thought of wearing flashy clothing can make him cringe. He usually chooses to wear dark tones and muted colors, and his wardrobe contains simple tailored clothing. As for accessories, a wrist watch or a simple gold chain might complement his outfit. This quiet person keeps a low profile and a distinct feeling of detachment permeates his aura. In a very stoic and silent way he works to protect his space. Even when walking with a group of people, he noticeably strives to maintain his privacy. If you make physical contact with him, his skin may actually feel cold.

An ideal location for his home would be in a country setting, without too many neighbors. Here he can have his need for quiet and privacy reflected in his surroundings. When decorating, this person will likely shy away from the idea of using warm honey tones. The painters can easily satisfy him, however, with various shades of slate containing hidden green undertones. He does not particularly like carpeting, but rather loves to walk across the floors of stone that one would find in an ancient cathedral. This individual makes it a point to conserve his energy as physical strength is not his strong point. Moving and heavy lifting are tasks that he should avoid. When engaging in such tasks, his body will frequently remind him that he was never destined to be a physical laborer. He owns a body that happens to need more rest than the average person, and it's evident that he tires easily. You may observe that he often asks others to bring him things or to turn on a light switch, simply because, for him, the act of getting out of his chair feels like work.

8 PHYSICAL POINTS

The expression, *He carries a lot of weight,* refers to the power a person has over others but can also describe the body of anyone with Eight physical points. This person conveys a powerful presence with his strong sturdy frame, and muscular build. I sometimes joke that if an Eight body type were hit by a truck, the truck would incur damage; but that person would walk away without a scratch. As weight is associated with power, it's not uncommon for the person with Eight physical points to be a bit heavy, and it can be nearly impossible for him to shed those extra pounds. Losing weight would be equivalent to losing power which is not an option in the realm of Eight. An intensely concentrated amount of strength within this individual provides more than enough energy and drive to fulfill the dream of becoming a professional athlete or a CEO.

The person with Eight physical points often uses his sense of touch to confirm that all of his selections radiate quality. When he goes shopping for clothing he may feel compelled to run his hands over the fabric to make sure the suit he is considering is fashioned from the very finest wool. Handling any product helps to assure him that it is constructed of superior materials. Whether furnishing his home or office, he painstakingly searches for exceptional pieces of furniture. His desk serves as his command center, so it is imperative that it be built of genuine hard wood and display impeccable craftsmanship. Some may call him demanding, but he is a person who knows what he wants and settles for nothing less than the best. He likes to be surrounded with quality, well fashioned, furnishings and imitation products simply do not interest him. He appreciates a good heirloom and makes each new purchase with the expectation that the item will last for many generations to come.

9 PHYSICAL POINTS

The person with Nine physical points likes to merge a little bit of heaven into her solid earthy existence. The choices she makes and the creative endeavors with which she involves herself are always inspired by the spiritual. She easily relates her physical being to the ethereal colors of the sky. This angelic individual enjoys playing with colors and effectively infuses it into every aspect of her life in order to make an impression on others. At times she feels most comfortable wearing pastels, pale blues, and every shade of white. Just when you think you have come to know her favorite color combinations, she will switch to the dramatic and intense hues of a rich sunset or the dark blast of an evening storm. This fascination with color keeps her experimenting with different shades of dye to highlight her hair. You may be surprised how quickly her tastes can change. Her hair may be flaming red today and then streaked with blondes and browns the next time you see her. She definitely has a flair for the dramatic and works earnestly to evoke reactions from others. Color rules as her most important decorating tool. You can see the effects of this in her house and furnishings. Although she can get quite theatrical, she always likes to keep her connected with the natural elements.

Typically, physical health is an interest of the person with Nine physical points. She likes to keep pace with the latest cutting edge theories for well being and is likely to own a yoga mat, a juicer, and have a well stocked vitamin cabinet. Alternative therapy treatments create a new paradigm of health care in her world. Nothing is ever commonplace with this person. Although she regularly connects with the spectacular, she will always tell you that it pleases her to live simply.

Emotional Expression

1 EMOTIONAL POINT

The person having only one point on the emotional scale is a rare find. He is not likely to be in touch with the array of feelings that most people take for granted. He is willing to briefly explore sensations and emotions he encounters concerning different issues, but he is not inclined to dwell on those feelings. Getting involved in deep emotional experiences simply doesn't interest him. Emotions are highly personal, internal experiences as far as he is concerned. With these boundries set in place, his inner disposition frequently leaves him unfamiliar with the feelings of others.

The web of emotional communication that most people share is lacking in his aura, and he routinely references experiences back to the mind for analysis. This reaction further closes his heart and leaves his feeling to be processed by the mind. Logic takes over as he examines the sensations sent from his emotional field. It does not go unnoticed that his egotistic resolve repeatedly motivates him to fulfill his own needs without concern for others. Consequently, many friends view him as being cold and inconsiderate.

When working with a person with One emotional point you must remember that he is innately self-focused. Narcissistic love becomes the primary target of his heart because his emotional communication has not yet evolved beyond that capacity. The limitations of sharing his heart with other people can be problematic, as he stays quite busy discovering ways to fulfill himself.

2 EMOTIONAL POINTS

The sensitivity of a person with Two emotional points should not to be underestimated. He truly has the ability to feel more profoundly than the rest of the population, and this delicate soul can be brought to tears very easily. On the positive side, he soaks up good feelings like a sponge. He experiences the total exhilaration of joy in the depths of his soul. Contact with negative emotions, however, can have equally profound (and sometimes devastating) consequences to his spirit. Unfortunately, this person readily picks up the abrasive energy left behind in a room where arguments have taken place. The world can be difficult, and constantly reacting to the negative energy that lingers in the environment can take its toll. A tendency to get caught up in the despairing emotional issues of other people exists in the person with Two emotional points. He must be forewarned that depression is a dangerous pit in which he can easily become lost. Hopefully, he has a strong mind to balance his overly sensitive emotions.

The person with Two emotional points should avoid watching news reports at all costs. Connecting with even slight sorrow can cause damage to his emotional body. I compare this profound emotional sensitivity to someone who has delicate skin. We are apt to remind the fair skinned person to wear sun screen and avoid sun bathing at midday. Similarly, we find the person with Two emotional points to be emotionally super sensitive. Exposure to negativity can, in a sense, burn his emotional body.

3 EMOTIONAL POINTS

The person in this emotional realm openly expresses childlike sentiments with a lot of hugs, smiles, and laughter. Three presents a very sweet, warm, and innocent expression of love. Openness easily draws people to her; however, don't be surprised if at times her childlike behavior can turn about and to become childish. This quick change is apt to occur when you least expect it. Picture the young toddler who we see laughing one minute and crying the next. Everyday events in the life of the person with three emotional points can take her (or him) from the heights of ecstasy to the depths of despair before finding her center again. Too many emotional upheavals can sometimes leave her feeling a little scattered; nevertheless, the energy of Three emotional points manages to help her land on her feet and re-engage her typical optimistic attitude. Although she has the capability of processing a wide range of emotions very quickly, she tends to stay in the positive end of the scale much of the time and usually appears happy as she ploughs through life's many challenges.

Being so honest and direct with her exchange of emotion allows her to easily relate to the feelings of other species. Animals sense and appreciate her wonderful kindness, and her ability to communicate with them. I offer one example of how this type innately understands the language of animals. I watched a friend who has Three emotional points get bit by a cat. This friend did not react with fear as many people would. Rather, she absorbed other messages from the cat through her emotional connection. Her response was to laugh and rub the cat's ears while she assured him that she respected his panther heritage.

4 EMOTIONAL POINTS

The person with Four emotional points feels the need to keep her emotions (both positive and negative) contained. In effect, self-control becomes her emotional priority. She is meticulously organized about how and when she will allow herself to express emotion. Consider the following two examples:

1. The woman with four emotional points attends a funeral for a co-worker. She may be as sad as the next person at the wake, however, she is also the one to look at her watch and announce that the allotted mourning period has ended and it is time to return to the office.

2. Instead of screaming with delight when she wins the lottery, this same woman, in a matter of fact fashion, will smile and call her accountant to have him handle her new assets.

Since she is so vigilant about restraining herself, others are apt to misjudge her. The control she exerts over her emotions should not be mistaken as a lack of sensitivity on her part. She has feelings, but the manner in which she handles her sentiments is very, very practical. Generally, a person with Four emotional points will be guarded about shedding tears in public because she doesn't want to appear weak to other people. As a result of her conscientious self control others tend to perceive her as being distant or aloof. Cold-heartedness becomes the errant perception of her, when in fact she may be a very caring person.

The person with Four emotional points feels safe expressing her emotions when she finds a secure setting. Her comfort level increases if her intimate relationship is allowed to develop in a conventional manner. She wants a diamond ring and a traditional, proper courtship.

5 EMOTIONAL POINTS

This swift current of emotional energy can move through the physical body like a streak of lightening. If a person with Five emotional points happens to win a prize, she will most likely scream, whoop, and do a little dance. When she has this type of excitement to process, she naturally feels the urge to jump up and down and hug the nearest human being.

When the Five heart wells up with love for her special sweetheart, she craves physical contact. Sexuality is her primary means of expressing her love. The extreme passion of her emotions fuels her drive to be a very sensual and erotic lover. This pleasure may be very exciting to her new boyfriend until he learns that he has also been selected to be the recipient of her wild emotional tantrums. When she becomes angry, she may feel the need to throw a few dishes at her boyfriend or smash his picture with a hammer. Five emotional points always require a physical release of passion. Anyone who carries Five emotional points should consider joining a gym and enrolling in an exercise class. These people need to burn off the excess emotional energy that charges through the body on a daily basis.

6 EMOTIONAL POINTS

The person with Six emotional points demonstrates a warm, loving and responsive disposition. He shares many of the same characteristics of those with Three emotional points, so you may want to review that section. As you do, consider that the person with Six emotional points represents a Three that has fully matured. The perceptive Six is equipped with a special internal feature that allows the bearer to sense and instantly respond to the feelings of other people. This feature enables him to easily communicate with others in a heart to heart fashion. When in doubt as to the needs of others, his solution is to send an extra dose of tender loving care to everyone around him.

The person with Six emotional points has such a generous heart that others usually find it difficult to keep up with his benevolence. There are times that this caring individual feels short changed because other people do not reciprocate with as much love as he radiates. He must understand that the majority of people do not enjoy the great emotional capacity he readily accesses. In order to keep things balanced, he may have to learn to keep his extreme generosity in check. This can be a frustrating and difficult task for the person who feels the constant urge to give of himself.

7 EMOTIONAL POINTS

She is the juror who straightforwardly announces, "He is guilty. Hang him."

The person with Seven emotional points can dish out a lot of tough love, and she does not wince or show any sorrow for the criminal who deserves the guilty verdict. Although many in her community perceive her to be cold and uncaring, she possesses very deep feelings. Her emotional exchanges are generally reserved for family members and the most intimate encounters.

Emotions are often associated with water and using this comparison we can surmise that the person with Seven emotional points has feelings that flow like a deep underground stream. Much of the time, this person may choose to keep her secret well capped and secure. It would be easy to walk right by and never notice the potential pool of refreshing water at your feet. This hidden and enchanting water is not accessible for everyone. Reserved for a few special individuals, only a privileged minority will ever be invited to drink the water. Emotions run deep in those with Seven emotional points. Those who do find their way into this heart will certainly be rewarded with a deep and loving relationship.

8 EMOTIONAL POINTS

This person is very businesslike and professional when it comes to expressing his feelings. To better understand the person with Eight emotional points I would recommend that you reread the information outlined in the section describing Four emotional points. Double the intensity of the Four energy and you will acquire a fairly accurate perception of the person with Eight emotional points. The control factor found in this category is extreme. The person with Eight emotional points becomes exceedingly concerned about restraining his emotions. He worries that others will perceive him as weak if he is caught crying or emoting in another way. There can be times when the iron will of the Eight breaks down. On those occasions, the person will express his powerful feelings and unleash a tremendous surge of emotional energy. The person with Eight emotional points is famous for expressing himself in an all or nothing fashion.

Since he can not conceive of doing things half heartedly, when he is ready to love, he does so with every fiber of his being. He regards traditions and social customs as badges of honor, and he adheres to them closely in matters of love. This individual is not one to waste time experimenting with feelings and he unhesitatingly makes the goal of his love clear. He enters into marriage prepared to accept all of the responsibilities enveloped in thos sacred vows. He regards the decision to have children as an expected part of this covenant. It can be quite disappointing if his partner does not agree with his strong feelings.

9 EMOTIONAL POINTS

Reflecting on the person with Nine emotional points, you become a witness to the ultimate drama. This person cycles through emotional highs and lows with incredible speed. It can be exhausting to watch him attempt to assimilate all the bits of data he collects through his feelings. The Nine vibration moves him to strive to resolve human suffering, and this becomes the special mission of his heart. He has compassion for every living creature and willingly opens his heart to all those in need. Pure intentions enable him to abundantly receive God's blessings, and the love of God becomes the fuel that allows him to continue on his magnanimous campaign of mercy.

Having an emotional center instilled with a desire to give keeps him faithful in practicing his charitable role. He has an affinity for working in crisis situations, and does not stop to analyze which causes are worthwhile and which are not. When a need is presented, his heart is moved to speed to the rescue. The generous energy of the person with Nine emotional points delineates the basic nature of the volunteer fireman or ambulance medic. This individual truly loves all and willingly serves all.

There are moments when the person with Nine emotional points may feel overcome with extreme loneliness, and he wishes someone would race to his rescue. When he falls into these low moods, he needs to rededicate himself to the common good. Ultimately, his devotion to service supplies him with the satisfaction he craves.

Spiritual Expression

The spiritual expression outlines the internal processes that uniquely connect each one of us with the Creator. We do not necessarily share this experience with one other as we do when we convey our thoughts and emotions. Spiritual advancement accounts for how each of us achieves communion with God. Some people refer to this communion as their union with the higher self or a higher power. Whatever we wish to name this connection is ultimately not important. The significance lies in building a strong relationship with the spiritual component of our being. In this way, we enhance our ability to clearly experience inner knowing (intuitive thinking). Although the spiritual expression remains private, we all certainly benefit when anyone opens to his spiritual potential.

People often tell me that they are unsure as to what being intuitive actually means. There is a popular distorted view that persons who read cards and tea leaves are the intuitive ones. I try to dispel this ridiculous belief by telling people that one of the most outstanding intuitive human beings we are all familiar with was Albert Einstein. During his lifetime he displayed psychic ability concerning physics and mathematics. People who easily express their talents, (knowingly or unknowingly) use their intuition to achieve success. Everyone has some degree of intuitive ability and it usually surfaces when they are doing the work they love, whether that is building bridges, cooking, teaching, writing etc. Intuition operates in a pure and simple way. You observe inner knowing functioning when you see a mother awaken in the night to check on her children.

We are spiritual beings and the intuitive awareness we enjoy blends with the thought process so seamlessly that it often goes unnoticed. When you fully accept the fact that our existence extends beyond the boundaries of the physical body, then it becomes a matter of logic to conclude that extrasensory power generates an effect within every person. This is our spiritual inheritance.

1 SPIRITUAL POINT

With One spiritual point the person understands that he is the center of the universe. By honoring the self, he honors God. He is not afraid to initiate progress and uncommon bravery frequently leads him to open doors to explore the unknown. Using his many creative talents he fulfills a leadership role and simultaneously finds unique ways to satisfy his own needs. Call him the spiritual trail blazer. Having One spiritual point is quite common in our culture as many spiritual explorers live on planet earth.

2 SPIRITUAL POINTS

If we begin with the premise that everyone has an antenna to God or their higher power, then it would be apropos to say that the person with Two spiritual points is using a satellite dish to make this connection. His communion with the creator is incredibly strong. Wherever he directs his attention, his intuitive powers become engaged. With little effort he is able to focus and absorb the answers to questions from a higher power. He displays superb instincts concerning other people and their situations; however, he is not inclined to share the information he is able to glean unless persuaded to do so.

3 SPIRITUAL POINTS

Language is the special gift of the person with Three spiritual points. Besides frequently conversing with his guardian angels, he feels completely at ease speaking to different classes of ordinary people. Somehow, he seems to find something to say about every issue and puts many at ease with his wit. He relates comic stories to others as a way of infusing humor into serious situations. The loose style in which he puts together information is highly creative. Listening to his manner of speaking reminds a person of the approach a young child takes as he playfully explores a new reality. The person with Three spiritual points speaks with ease.

4 SPIRITUAL POINTS

The person with Four spiritual points refers all of his spiritual experiences to the left brain. He strives to make his finest encounters with the divine tangible and useful. For example, after receiving a message in meditation, he may feel the need to record the experience. He will carefully file this information so that he will have ready access to it should he choose to write about the experience in the future. He tries to make sense of everything and frequently asks the Creator, *"What is the purpose of life?"* He has intelligent hands and will often find himself tackling projects and fixing things without any prior study or training.

5 SPIRITUAL POINTS

The person with Five spiritual points connects with knowledge as if he were a living encyclopedia. He seems to have an especially powerful magnetic force that attracts random data. The constant flow of information moving through his aura can be quite overwhelming.

He routinely draws psychic energy into his physical body. This is the type of psychic reader who gets messages for his subjects by holding a piece of their jewelry in his hand. Because of his natural talent to extend his feeling, he can become ill when walking into a room where a lot of negative energy has been exchanged. It's important that he realize the consequences of his great sensitivity and avoid such encounters.

6 SPIRITUAL POINTS

Anyone with Six spiritual points *knows* the concept of soul mates to be true. His high self moves him to honor marriage as an expression of Divine Love. Quite naturally, this person demonstrates to the world what it means to be an exemplary parent. Communication with the spiritual realms provides this person with an intuitive means of choosing the best forms, textures, and colors to balance any work of art.

7 SPIRITUAL POINTS

The person who has Seven spiritual points fully realizes he is a holy spiritual being and he regards others as having the same status. He possesses a direct connection to the wisdom of the ages. Although he has access to the akashic records, he is not apt to share this awareness with others. Spirituality is a very private matter for him.

8 SPIRITUAL POINTS

This is rarely found. An individual with Eight spiritual

points thoroughly understands the mechanics of life. He is able to easily transcend the material plane and move into other dimensions at will. He has the capacity to use his astral body to explore the cosmos. Understandably, it is difficult for him to share his awareness with the common man.

9 SPIRITUAL POINTS

This is rarely found. With Nine spiritual points, a person relates to the energy of the Christ Consciousness. He opens his heart in service to all people. He understands the divine grace of compassion to be the ultimate prescription for healing all human suffering.

Conclusion

The language of numbers is an important key that has helped many people to unlock secrets held inside themselves and others. For ages philosophers have used these keys called numbers to transform their lives. Today, the knowledge continues to provide people with a unique perspective of themselves and their relationships in this wonderful, creative universe.

It is my hope this journey has allowed you to develop greater awareness of yourself and others by applying the new insights you have gained. Allot sufficient time to ponder the creative forces. If you give it a chance, the intelligence inherent in numerology will begin to rewire a new format of understanding into your thought patterns. After going through this process of acceptance, whenever the number Three presents itself, the playful creative child will be right there with you. The integration of the language of numbers into your thinking will help you to appreciate the fact that the number and the energy are the same.

The section of the book on applications was purposefully limited. We reviewed the following areas:

1. The cornerstone of the name
2. The Karmic Table
3. The Expression of self (mental, physical, emotional, and spiritual)

I chose these applications expressly because the calculations are simple. I did not want you translating your whole name into numbers and then adding up the numbers and reducing that result into one solitary number.

As you continue to work with me, you will find that I do not endorse reducing every number as you calculate a chart. I stay away from this method whenever possible because experience has taught me that reducing numbers can be misleading. The technique oversimplifies this marvelously rich language and can leave a person with a limited understanding as to the power of numerology. Too often when entertaining at events I have had people say to me, "Oh, you do numerology. I know what that is. You add the numbers up in your name. I already know that I am a number Three. Could you please do something else with me?"

I am always dumbfounded when I encounter this attitude. As I proceed to work with the person, I make sure to enrich his or her information base so they can experience how powerful the language of numbers can be. The feedback I frequently receive about me is: "You cannot know these facts from numbers."

The little bit of superficial information that the person initially acquired concerning numerology has clearly misled them. Learning the truth about the subject matter propels him or her into a state of becoming overwhelmed. Misunderstanding keeps many people from acknowledging the validity of numerology as a humanistic science.

The express purpose of this book is to ground you in the basics of numerology and arouse further interest in the subject. My objective in writing this book was to formulate this knowledge into a simple uncluttered format, so you would have a reference that delivers clear and accurate number insights. This material should help prevent you from falling into the trap of memorizing a lot of data concerning each number. It is my intention that you truly own this knowledge. This is my gift to you.

I invite you to continue the journey with my next book ***Double Digit Numerology.*** In that sequel I will explain the importance of the Place Value of Numbers. When you learn the simple process of reading combinations of numbers, I expect that you will never be satisfied with the limits of single digit numerology.

Numerology, the Language of Numbers is the key. I encourage you to turn the key by continuing to study ***Double Digit Numerology*** with me.

Numbers in Motion

1
Leaping

2
Walking

3
Skipping

4
Marching

5
Running

6
Dancing

7
Resting

8
Jogging

9
Gliding

Best Wishes with your new knowledge of

Numerology: Key to the Language of Numbers

About the Author

Fran Szarnicki is an experienced reader, teacher, and lecturer in the field of numerology. She applies her intuition to interpreting the mystical codes of numbers. Through her personal consultations, she has helped countless people to accept themselves, recognize their own abilities, and expand their creative potential. Using the language of numbers, she prepares people to take control of scheduled opportunities destined to emerge in their lives.

Numerology: Guide to Understanding the language of Numbers is the first in a series of books covering the philosophy of numerology, its applications, and practical uses.

She is currently working on her second book in the series: *Double Digit Numerology*. She writes, lectures and teaches classes in The Villages, Florida. There is more information on her website, www.NumerologyWizard.com.

Made in United States
North Haven, CT
10 November 2026

22088877R00124